The Speed Math Bible

by Danilo Lapegna

Master Mental Math, become a better problem solver, conquer your daily challenges.

Your best "Return on Investment" ever!	1
II - 5 strategies to Improve Your Life through mathematics	**8**
Play "Quantitative Modeling"	10
Measure where no one else would	11
Improve the quality of your daily "measurement tools"	12
Where numbers are unforgiving, make them your allies	14
Think more often in terms of cost-benefit	15
III - Three Fundamental Steps	**18**
1 - Keep the right attitude	18
2 - Make your memory your best ally	20
3 - Strengthen and Enhance Your Foundations	25
The Fundamental Properties of Arithmetic Operations	26
Dividing or multiplying by 10, 100, 1000, and other powers of 10	28
Addition and multiplication tables from 0 to 9	30
Addition table	31
Multiplication table	31
IV - Instantly Multiply with Your Fingers	**34**
V - Mathematics and Hindu Wisdom	**40**
VI - Composing, Decomposing, and BlackJack	**46**
Searching for tens and squares	46
Borrowing technique	48
Break that down	49
VII - The Magic Column	**54**
VIII - Introduction to Game Theory	**60**
IX - Break, Simplify, and Save Money	**69**
Addends decomposition	70
Factor decomposition	72
Expressions decomposition	75
X - The World of "Golden Numbers"	**82**
φ	82

3 84
9 86
11 89
37 92
143 92
666 93
1089 94
2025, 3025, 9801 95
3367 95
6174 95
37.037 96
142.857 97
12,345,679 97
1,016,949,152,542,372,881,355,932,203,389,830,508,474,576,271,186,440,677,966 97

XI - Rapid Multiplication from Hell 100

Multiplication by 11 103
Multiplication by 12 105
Multiplication by 6 106
Multiplication by 9 107
Multiplication by 5 109
Multiplication by 7 110
Multiplication by 4 111

XII - Vertically and Crosswise 115

XIII - A wonderful Connection 119

Two-digit numbers in which either the tens or the units are the same, while the others add up to 10. 119
Numbers between 11 and 19 122
Numbers that are "equally distant" from any integer 123
Numbers close to a power of 10 124
A summary and some exercises. 128

XIV - Chinese Graphical Multiplication 131

XV - The Sliding Cross 139
XVI - Travel, Discounts, and Elections 151
Rounding, nihilism, and "catastrophic" errors 154
"Killing" the decimal point... and expressions decomposition! 157
Addends decomposition 159
Percentage decomposition 162
Now use these… to perform quick multiplication without decimals! 164
The traveler's decimals 165
How to triple your capital 167
Let's defeat one last "monster" 169
XVII - A Game of Chances 176
Expected value, and two ways to predict the future 178
Dice, coins and random stuff 181
Negations, coincidences, and mutual exclusions 182
Casino and multiple choices 183
When to insist on "Spinning the drum"? 188
XVIII - Estimation and Casting Out Nines 193
Estimate: rounding with zeros 194
Estimation: replacement with fractions 196
Casting out nines for addition and subtraction 197
Casting out nines for multiplication 199
Casting out nines for division 200
XIX - 10 strategies for calculating Squares... and one to save your life! 204
1 - Learning Basic Squares 206
2 - Exploit the technique of multiplying numbers between 11 and 19 207
3 - Strategy for numbers ending in "1" 207
4 - Strategy for numbers ending in "5" 208
5 - Strategy for numbers ending in "25" 209
6 - Strategy for numbers starting with "5" 210
7 - Reuse the technique for multiplying numbers "equally distant from an integer." ... 211

    8 - Take advantage of the proximity to another square 212

    9 - Reuse mental multiplication techniques for two and three-digit numbers. 213

    10 - Reuse the multiplication technique for numbers close to a power of 10 215

XX - Square root, cube root, and... thirteenth root! 220

    How to identify a non-perfect square. 221

    Two methods to quickly approximate Square Root 221

    Tricks for Cube Root and Fifth Root 223

    Thirteenth Root? 225

XXI - Mathematical tools 229

    Estimate, with Gauss 229

    When to say "Yes"? 231

    Ideal weight (or almost) 232

    Hypothetical Close Encounters 233

    Cosmic Card Decks 234

    Save and feel better, with the 33% Rule 236

    Managing chaos 237

Conclusions 240

But there's more... 241

The author 248

Bibliography and Further Reading 249

Appendix: Table of Prime Numbers from 2 to 5000 250

Disclaimer 254

This isn't an AI-generated book, and that's by design.

Yes, we use cutting-edge technology to polish our writing and augment our research, but **the heart of this book is all human:** painstaking research, deliberate crafting, and *a lot* of late nights! Just so that you know: **what you've got here has been shaped with heart, intention and care!** Oh, and also, in case you don't like something: you can blame us, not the robots!

Your best "Return on Investment" ever!

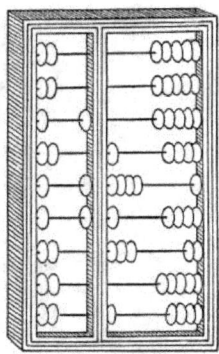

An increasing number of professors, scholars, and researchers, such as Prof. Jo Boaler from Stanford University, Dr. Keith Devlin, known as "The Math Guy" on NPR, and Prof. Carol Dweck, author of "Mindset: The New Psychology of Success," now agree that the traditional school method of teaching mathematics—whether arithmetic, algebra, or calculus—has many structural problems. According to them, and many others in the field of mathematics education, like Prof. Paul Lockhart, author of "A Mathematician's Lament," and Prof. Conrad Wolfram, founder of Computer-Based Math, this method, in most cases, does not help students navigate the subject in a way that is interesting, in-depth, or valuable for facing the real world.

If I had to, specifically, summarize the issues that clearly seem to emerge from our mathematics education, I might say that:

Individual creativity is encouraged far too little: in school, there is often a tendency to teach that "this is the way and that's it," and as such, everything must always be calculated the same way, without ever questioning the pre-established procedure. It's a system that, to use a perhaps clichéd but undoubtedly true phrase, provides too many answers instead of teaching students to ask questions. This approach can only produce automatons in those who choose to "embrace" the mechanism, and complete alienation in those who understandably reject it.

Precisely because I am aware of the enormous transformative and educational power of asking the right questions, I decided to design this book with the aim of completely dismantling this closed and limiting perspective of the subject. I want to do everything possible to transform it into a genuinely strategic challenge, one that's profoundly creative, intellectually stimulating, and, I dare say, often incredibly fun for this reason.

The concept of a mathematical "trick" is often demonized (or at least, given little importance): as mentioned previously, "standard" mathematical procedures are presented as the only possible path, concealing from common knowledge the truth that various types of arithmetic, algebraic, or analytical operations can be calculated through actual "strategies"; or, as I like to call them, "hacks" that, to the less purist, might seem like "cheap tricks," but in reality, are simply quick, effective, and often intriguing shortcuts to reach the solution. Of course, adopting shortcuts always has its own list of pros and cons, but the point here is not so much to examine all the practical or ethical consequences of thinking in terms of "strategies," but rather to offer those who study the subject a rich array of tools to practice problem-solving strategies that consider the evaluation of connections, possibilities, and alternatives, rather than uncritically adopting some predetermined "single path."

No attention is given to the importance of individual expression: too many people feel a true sense of "alienation" towards certain arithmetic concepts because the rigid methodology discussed prevents everyone from expressing themselves according to their own essence, tastes, and natural inclinations. This is where the approach of this book will come into play once again, deeply oriented towards personal expression and the acquisition of multiple calculation strategies, enabling you to operate each time in the way that is most comfortable, congenial, and... truly yours!

There is little to no discussion about the immense practical utility of mathematics, both in competitive terms and in understanding the world: beyond the ability to calculate your restaurant bill more quickly, the enormous added value you will have in any competition or test, or the potential development of your brain areas associated with reasoning and working memory (as suggested by studies like "The Impact of Achievements in Mathematics on Cognitive Ability in Primary School" (2022), or "Neuroscience of Learning Arithmetic—Evidence from Brain Imaging Studies" (2009); which in itself, I would say, offers absolutely enormous value), in this book you will discover that mathematics can provide you with many practical tools to better " navigate" the various domains of everyday life.

Mathematics, as Galileo stated, is the primary alphabet of the physical laws that govern our reality, the evolution of phenomena, and even the uncertainty that characterizes them. Therefore, aligning ourselves further with these laws, understanding and internalizing them, can also help us better comprehend this reality and, in theory, become more effective within it, providing us with the foundation to acquire a more deeply scientific, factual knowledge, distant from thoughts or beliefs of a "magical" nature.

It is also worth highlighting how mathematics has become, by an obvious transitive property, the essential fuel of every technological marvel in our contemporary world. Computers and tablets would never have come into existence if two mathematicians like John Von Neumann and Alan Turing had not laid the mathematical foundations of the first computers. The internet would never have been possible if no one had developed the mathematical principles underlying network theory. And search engines and AI could never have been created without the equations, algorithms, and methods that allow for systematic ordering and classification of any chaotic data set. It's no coincidence that the inventors of Google and modern-day revolutionaries, Larry Page and Sergey Brin, both hold degrees in mathematics. I'm not saying that every reader of this book will end up founding the next Google, but I'm absolutely certain that greater mathematical proficiency in a world running on these frequencies can help each of us be more of a protagonist and less of a spectator in the complexities of our world.

Finally, it is crucial to highlight the "indissoluble" relationship between mathematical knowledge and economic-financial competence. This not only refers to the importance of mathematics in developing the ability to plan investments wisely but also to how it can positively influence the management of daily expenses. I'm sure this is of significant interest to everyone, considering how central the "money management" factor is in our lives, often representing one of the most relevant and impactful sources of daily stress.

So the evidence seems to suggest that every single minute spent improving the quality of your skills and mathematical knowledge remains one of the best returns on investment you can ever make. And so, with the intent of particularly reaching the hearts of the curious, "intellectual experimenters," and also parents who want to acquire efficient and fun tools to "mathematically educate" their children, I have worked for months to compile this manual. The book you are reading at this moment represents, probably, one of the most comprehensive collections you may ever find on the topic of instant calculation strategies, the "philosophy" of mathematics, tales of "beauty" in the numerical world, snippets of history, and everything fascinating, interesting, and useful that can be found in this vast universe. Every section of the book, besides analyzing different aspects of the subject, is accompanied by a set of stories, philosophical connotations, practical applications, both daily and playful, of everything covered up to that point. Moreover, for all those who wish to delve further into the functional and "daily rewarding" side of this discipline, I have prepared an additional section of "useful formulas for everyday life" at the end of the book. Feel free to browse through it whenever you have the desire to discover some interesting "life hack" and practical tool to better tackle and plan your days.

Author's Final Note: Since many of you might start from a position of "substantial hostility" towards the subject, I would encourage you to approach this book with a mindset that allows you to take all the time necessary to "digest," understand, and learn to apply the information contained page by page. Don't focus too much on "reaching the end," but try to enjoy the journey, making it a gym for the mind. The real goal, in fact, can never be in the "end" of the book or a chapter, but in the journey you outline for yourself, and in gaining ever more knowledge and competence from that journey. So stop, reflect, experiment, and try to apply what you read in everyday life. Put your book away for a few days if you feel the need to take a break, and pick it up again when you feel like learning something new. **Remember that commitment will be needed;** then, make this volume a friend, a valuable ally, and a companion on your journey of intellectual growth.

But having reached this point, I can only conclude the introductory section by wishing that this reading brings you the best, whether it's a desire for learning, a mental challenge, or simply the satisfaction of intellectual curiosity. Good luck!

At the "Kintsugi Project," we celebrate the joy of reading every day. If you've got a physical copy of this book, share the love! Snap a photo and tag our Instagram account, @danilolapegna.kintsugi, using the hashtag #kintsugibooklove. We'll be delighted to send you a personal thank-you.

For feedback, proposals, requests, or suggestions, don't hesitate to reach out to us at info@kintsugiproject.net. We're always happy to hear from you!

Danilo Lapegna

- The Speed Math Bible -

- The Speed Math Bible -

II - 5 strategies to Improve Your Life through mathematics

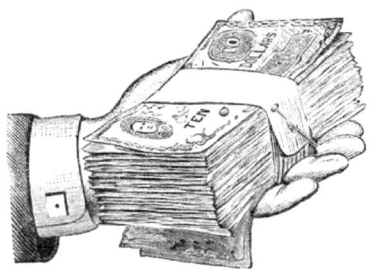

The act of counting and measuring is one that exists in harmony with the most common tasks of our daily lives. How many hours of sleep are needed to feel fully rested, our optimal body fat percentage, the time required to complete those errands, the number of kilometers before our fuel tank is empty, are all examples of questions that inevitably arise in our daily routines and theoretically require optimal and effective solutions.

After all, each of you should be well aware that you live within a vast system literally governed by mathematics: as mentioned several times in the introduction, we are "inevitably" part of a universe where every aspect, from physical magnitudes to economic variables, is in constant flux, and every fluctuation is consistently regulated by intrinsic mathematical laws.

In the same way, however, we too often forget this reality, and therefore the importance of reasoning more frequently from a quantitative perspective; even though we are indeed immersed in a universe of a mathematical nature, our brains do not seem structured to easily or immediately tackle problems in a quantitative manner, often reasoning more in terms of ideas and meanings rather than numbers or quantities. After all, our intellect evolved predominantly in a prehistoric context where formal mathematics did not exist; and as you can easily imagine, these contexts were often extremely hostile and constantly generated situations where it was necessary to make quick decisions based not on precise calculations, but on very general and intuitive interpretations of the surrounding world.

And so, even today, this affects our innate ability to read the world and consequently plan solutions to our problems; for instance, when we have to make complex decisions, such as economic or strategic ones, we tend to rely on opinions, intuitions, or "impulsively suggested" truths rather than detailed quantitative analyses. But most importantly, we will too easily swing, in real-life situations, from one extreme to another: not always being ready to negotiate with the world's "quantitative" complexities, we often reason as if we were "living switches," machines that only understand "0s" and "1s," "yes or no," "on or off"; and thus, we end up adopting solutions, joining factions, supporting ideas because they are "good" or "bad," not because we consider their nuances, the necessary gradations of meaning, the possible costs compared to the benefits, and how these impact the overall context. Hence, in avoiding these complexities, it is likely we will live with continual disappointments, violent emotions, "ruinous" shifts between life phases where we seem to lack any real "center of gravity" to hold onto.

Learning to reason more from a quantitative perspective, and therefore learning to more frequently measure, evaluate, and calculate the quantities involved, while adapting to nuances and gradations, can provide both greater efficiency, and thus a higher likelihood of achieving more with less, as well as a better overall management of the complexities we will inevitably face, whether they are economic, emotional, or affective.

What has been said so far, let's be careful, does not mean crossing the line into an obsession with rationalization or measuring every aspect of reality. Nor does it mean that every moment of our existence should be subjected to endless experiments, data collection, and calculations: the risk of adopting such a mindset, at best, might be to imprison ourselves in a perpetual state of analysis paralysis, where everything is constant computation, and we are unable to achieve anything concrete. Intuitive, irrational decisions; choices made with a mindset of "pure" experimentation and improvisation are absolutely crucial to leading a life that is healthy, balanced, and "intelligently productive." However, the fact remains that sometimes, when there is a clear need to achieve better results in a certain area, applying a firmer "mathematical grip" on it, understanding which quantities are involved, measuring them, evaluating the results, and acting based on them, can much more easily lead us to achieve what we desire.

After all, if we look at any major success, particularly in the economic and business fields, none of them (barring perhaps very rare exceptions) occurred as a result of instinctive or random decision-making. Instead, every great product or service has been the result of "winning" decisions made by true "measurement professionals." Behind many of the things we enjoy with satisfaction every day, there has been a whole process of measurement and evaluation on factors such as the potential returns of market niches, production costs, and all those KPIs (for those who may not know, Key Performance Indicators) deemed relevant for determining whether or not significant objectives are being approached.

But how could we effectively work to improve the quality of our thoughts, our actions (and… our life?) through the application of these principles? Here are some guidelines and a few mental "exercises" that might prove extremely useful:

Play "Quantitative Modeling"

The concept suggested by the title is quite important, as we will revisit it later when discussing how to analyze situations through game theory and probability. For now, however, try to make it your own from a more general perspective by treating it as a "kind of game" to apply to any situation in your real life. For instance, if there are three actions that make you feel particularly good, instead of simply thinking of them as "good" or "optimal," try estimating the amount of benefit you feel from each. You could assign each a score from one to ten, or one to a hundred. When you have a number to work with, the information, although possibly simplified by the modeling process, becomes easier to process and compare to similar information. For example, after giving a score to each of the "three things you love to do most," you might realize that one of them is worth much more to you than the others, prompting you to focus on it for a greater amount of time. Alternatively, you might see that each of them gratifies you equally, leading you to decide to divide your time among them evenly. The important thing here, however, is not so much to focus on the action to take as a consequence, but rather to "train" yourself more often to think in terms of the quantities involved in our daily processes.

Measure where no one else would.

It is said, according to a maxim erroneously attributed to Einstein, that "Not everything that can be counted truly counts, and not everything that truly counts can be counted." This principle, despite its "dubious" origin, not only tends to be true, confirming a "bitter truth" about the inevitable limitations of mathematical and numerical models, but also makes us realize that too often the most commonly used measurement systems might divert us from what it would be more useful to focus on.

An example I love to use regarding this topic is that of a "report card" given to a student halfway through the year. We often tend to see it as a measurement of a student's effort, without forgetting that on the other side there is a teacher with their own limitations, biases, and possible methodological errors. And this is precisely where the small revolution of "measuring what no one else would measure" could come into play: an evaluation of the quality and worth of the teacher's work by the students, for instance. This would help to understand the true origin of the problem and, consequently, to find a much more effective solution.

This is just a simple example, perhaps even challenging to apply, but the general principle to be drawn is: measuring, evaluating, and focusing where others tend not to are all things that could give you a real additional "competitive edge." Try to understand where this concept can be applied in your daily challenges, and it could provide you with interesting insights to potentially spark even small conceptual revolutions.

Improve the quality of your daily "measurement tools"

If we were to creatively merge two sayings like "What you measure defines your field of attention" and "Where you focus your attention, there you achieve results," we might say "What you measure, and how you do it, defines, at least in part, the scope and quality of results." Let's imagine that our goal is to reduce body fat, increase muscle mass, and maximize physical performance, and let's explore how applying this principle can help us achieve our goal more quickly.

For example, you might consider pursuing your goal based on the information from a single nutrition book or a single workout plan, deciding based on which one promises quicker results on the scale (a "single" or otherwise limited measuring tool). However, you would certainly have better chances of achieving satisfactory progress if you increased the number of your measuring tools, as would happen if you consulted and compared the advice of multiple fitness and nutrition experts, or if you used various tools to monitor your progress, such as body fat analyzers, fitness apps, and bracelets that track your vital parameters.

In this context, where one often deals with multiple data points from similar sources or from the same source, the concept of the *mathematical mean* can be particularly useful. This concept is known to most people, but it is worth "refreshing." If you have multiple quantities, measurements, or values related to aggregable, consistent data (clearly, avoiding the classic "mixing apples and oranges" as taught in school), you could try summing them together and then dividing the result by the number of values you've added. This gives you the mean of the measurements, a highly valuable figure because, to simplify, both measurement theory and probability theory suggest that it often tends to approximate the "real" value, if such a thing can be discussed in the given context.

Let's take, for example, the measurement of your body fat index. Suppose you measure it every week for a month and obtain the following results: 20%, 19%, 18%, and 20%. The mean here would be (20 + 19 + 18 + 20) / 4 = 19.25%, and this can give you a general idea of your progress, despite the "inevitable" weekly fluctuations.

But in addition to the mean, there are two other mathematical concepts that might be useful to you: variance and mode. Simplifying, variance indicates how far your measurements deviate from the mean. For example, if your body fat percentage varies a lot from week to week, you'll have high variance, which could suggest that you should try to stabilize your diet or training regimen more. On the other hand, mode is the value that appears most frequently in your data set. If you've measured your body fat over a year and the most common value was 18%, this would be the mode; a value that could give you an indication of a possible "state you tend to stabilize towards."

But what can truly add value to your measurement is the enhancement of your tools' quality. How many times, for example, has a highly praised training plan left you exhausted without delivering the expected results? Seek "better," "gather higher quality data," and you might achieve your goal in a "surprisingly" easy way. Try, for instance, to filter only the data and advice that have greater scientific relevance, or, to improve your action plans, rely only on the experiences of people with a similar body structure and physiology (assuming you can assess this with sufficient accuracy), and even a smaller amount of data might prove to be much more accurate, reliable, and effective.

It's clear that, as someone here might have already noticed, these are oversimplifications of typical concepts in measurement theory and data science, the art of combining numbers and data to extract conclusions, action plans, and potential predictions. But once again, let's leave the details to the professionals and try to take these simplifications for what they are: small conceptual gateways to the world we're trying to introduce you to— a world where greater attention to the accuracy of the measurements of the quantities surrounding us can become the "master key" for better actions, better thinking, and better planning.

Where numbers are unforgiving, make them your allies.

There are and will always be fields of action, contexts, and situations governed by "hard" mathematical laws, where variations in quantity have predominant importance and cannot be reconsidered or avoided. On the other hand, we have just talked about taking measurements and rigorous data "seriously" in our action plans only if the situation truly requires it; and perhaps here lies one of the fundamental principles that defines the significance and "seriousness" of this requirement: in the rigor of the laws that govern this situation. Any decision that does not adhere to such preliminary measurements will prove unfruitful and counterproductive.

It's like managing our personal finances: if we spend more than we earn, we'll end up in a critical financial situation, and no qualitative, intuitive, or empirical factors can compensate for this reality. The same applies to building a piece of furniture for our living room: if the wood is designed to support less weight than the object you'll place on it, your furniture will break, whether you like it or not. It may seem trivial, but often these factors are the first to be overlooked, with all the negative consequences that can follow. Therefore, making extra measurements when necessary, actively tracking your monthly expenses, calculating the weight capacity of that wooden plank, or ensuring that you stick to the long-term budget for that project are necessary tools not only for aiming for superior quality but also for avoiding potential disasters.

Think more often in terms of cost-benefit.

As mentioned a few pages back, the cost/benefit ratio is probably the "ultimate" tool in decision theory; it is the one tool that perhaps more than any other helps to regain an objective perspective on things and set aside preconceptions, biases, and cognitive distortions. Imagine, for instance, that you have several job offers, each with its own advantages and disadvantages, and you aren't sure which one to accept. Or perhaps you are house hunting, have multiple options, and want to determine which is the most beneficial. To simplify the situation and figure out the most advantageous choice, you could rely on the option with the highest cost/benefit ratio. In other words, for each option, list the disadvantages it presents and rate each on a scale of "harmfulness" from 1 to 10 (or from 1 to 100).

So, for each option, calculate the mean of the scores for its disadvantages and call it the "Harmfulness Index" of that option (ensuring that this index is never zero, and therefore at least one "disadvantage" must be different from zero).

Then, for each option, make a list of the advantages, assigning a score to each one. After that, do the same as you did with the disadvantages, calculating the average score of the advantages for each option. Call this mean the "Advantage Index" of the option. Make sure to use the same scale for scoring: if you rated the disadvantages from 1 to 10, then also rate the advantages from 1 to 10. If you rated from 1 to 100, use a 1 to 100 scale for the advantages as well, and so on.

Now, for every possible decision, divide the advantage index by the harm index. This will be the cost/benefit ratio of that choice, and you will likely notice that the most advantageous option to take will be the one with the highest ratio.

Clearly, since in this case the assessment is based on "ratings" rather than strict measurements, it will be purely indicative.

However, it serves well as an additional "sufficiently approximate" tool to make decision-making more efficient and meaningful. Of course, in all those cases where greater rigor is needed, we will act accordingly.

Furthermore, it might be beneficial to emphasize that if there is a certain degree of uncertainty in the factors evaluated as advantageous or disadvantageous, it's always better to base your decisions on other types of indicators, such as those we will discuss in the chapter dedicated to probability theory.

The Historical Pill

It can be assumed that mathematics was born alongside the very necessity of humans to count and measure. The earliest signs of "mathematics," if we can call it that, date back to the Paleolithic era, when humans began counting stones, marking quantities of things deemed useful to track, such as supplies, stars, days, and probably even enemies. However, true mathematics as we know it today likely had its real origins around 5000 years ago, a time when humanity began settling in large agricultural communities, and the first kingdoms and cities emerged. In this new social reality, the art of calculation became essential for activities like managing crops, constructing buildings, taxation, and astronomy. The earliest evidence of a real "mathematical art" comes from ancient cultures like the Mesopotamian, Egyptian, and Indian civilizations. Sumerian clay tablets, for example, already document the use of highly complex numerical systems, while the Egyptian pyramids themselves testify to a remarkable understanding of geometry.

The Aphorism

"The engine of mathematical invention is not reason but imagination."
(A. De Morgan)

Mathematical Oddities

Benford's law is a "surprising" statistical phenomenon that appears in many sets of natural data: the list of the world's longest rivers, city populations, earthquake data, and many others. This law states that in many of these datasets, the initial digit "1" appears as the first digit much more often than one would expect if the digits were uniformly distributed. Specifically, "1" tends to be the first digit about 30% of the time. Furthermore, this phenomenon can be used to detect fraud in financial or electoral data, as "artificially fabricated" numbers often do not follow Benford's law.

III - Three Fundamental Steps

I will now begin by explaining three fundamental factors that you can start working on to improve the accuracy and speed of your calculation skills, and consequently, all the strategic tools that will rely on them. These are somewhat like those essential "quid" that should be nurtured even before learning the advanced calculation strategies that will be discussed later. By themselves, they will already significantly help enhance your mathematical reasoning, both numerical and beyond.

1 - Keep the right attitude

The "pure attitude" is an element as undervalued as it is extraordinarily powerful when it comes to potential and performance. If, for example, you keep telling yourself, even subconsciously, that math isn't for you, that certain calculations are "too difficult," and that all of this "will be useless," you will surely make no progress (and yes, self-sabotage is an extremely complex topic that can apply to our attempts to progress in any field). This, of course, doesn't mean that "telling ourselves we are geniuses" will magically open an infinite number of doors for us. However:

Remember that whatever you believe to be your innate predisposition, you are under no obligation to adhere to it. Yes, genetic predispositions exist; however, it would be foolish and extremely reductive to believe that our genetics "represent the whole story." Just think of all the practical examples that contradict the "determinism" of this principle: Gert Mittring, for instance, who won the gold medal for nine consecutive years in mental arithmetic at the "Mental Skills Olympics," was one of the worst in his class in mathematics at school. And before him, many other "brilliant minds" we have evidence of (what that actually means is another discussion entirely, but let's save that for another time), like Thomas Edison, faced extreme difficulties in "blooming" within the norms of formal education.

In short, not only are our past results not an indication of our true aptitude for a subject, but it is ingrained in the biological, plastic nature of our brain that we are always the ones who choose which parts of our potential intelligence to cultivate and which to let wither and decline. Therefore, I would bet any amount you wish that if you feel "not inclined" towards mathematics or calculation, it's because there was *also* a lack on your part in terms of perseverance, or on your teacher's part regarding the learning methods conveyed. And by the end of this book, if you diligently apply the strategies I will illustrate, you will have developed skills you never even imagined you possessed.

Exercise your mind. Try to use your smartphone less and practice doing everyday calculations in your head more often, like figuring out the bill at a restaurant or the cost of groceries at the supermarket. If you're not motivated by the "pure intrinsic pleasure" of improving a mental skill, try to do it whenever it's related to an activity you care about. For example, if you are very invested in your physical fitness and want to monitor your daily calorie intake, don't always rely on the latest app; instead, try doing the necessary calculations in your mind. Understand how many grams of protein, carbohydrates, and fats you're consuming and how these translate into calories. Or, if travel is your passion and you're planning a grand intercontinental adventure, try doing currency conversion in your head before checking the app on your phone. In its very simplicity, this advice alone will greatly increase your "mental math readiness," allowing you over time to develop more efficiency, precision, and speed not only in your calculations but in your cognitive processes in general. Given the extraordinary plasticity of our brains and how mental math intensely challenges our concentration and working memory, continuously practicing and thus strengthening the neural circuits associated with these activities can greatly improve our reasoning abilities and focus, as well as short-term memory use. As has been well known since the time of Aristotle: "We are what we repeatedly do. Excellence is not achieved through a single action, but through the establishment of a habit."

Don't be too stubborn. Yes, challenge your abilities, and sometimes aim to exceed what you previously saw as a limit, but don't "stubbornly persist" on problems you just can't solve: sometimes the best solutions come when you take a break and start doing or thinking about something completely different! In fact, even when we abandon the process of solving a problem, our brain often continues to work "in the background," providing us with the best ideas literally "all of a sudden," and mathematical problems are no exception to this rule. Famous, after all, was the expression "Eureka" which Archimedes, according to legend, exclaimed after finding the solution to the problem of calculating the volume of solid objects while relaxing in a bath (and not while scribbling with pens and parchment). Legend or not, in any case, when the solution to a problem doesn't come, whether mathematical or of any other kind, try taking a break and moving on to something else. Forcing yourself to postpone working on a problem helps you approach it from perspectives you might not have thought of, and although it may seem counterintuitive, it can be the best way to help your brain do its job.

2 - Make your memory your best ally

Any calculation method essentially consists of two active phases: one is the actual calculation, while the other involves storing the previously obtained results, both partial and total. And this latter phase, as mentioned earlier, is probably the most critical of all.

Consider the case where you want to solve a two-digit column multiplication mentally, without the aid of paper and pen: what will likely be most challenging for you is not so much the calculation itself, but rather the need to repeatedly memorize the numbers produced by the single-digit multiplications required to reach the final result.

For this reason, on one hand, as you will see, many rapid mental calculation strategies have been devised in such a way that the user needs to minimize the elements to be memorized. On the other hand, training one's short-term memory becomes a crucial element for those who want to enhance their arithmetic skills.

So try following these tips, and monitor which of them prove to be most effective in improving your problem-solving skills (I would say, both mathematical and otherwise):

Focus. It may seem trivial, yet absolute focus on what you are doing greatly enhances your ability to memorize its components. Try to isolate yourself, as much as possible, from sources of distraction and learn to concentrate all your mental resources on what you are processing at that moment. As mentioned earlier, the more we strive to do this, the more we will strengthen the neural circuits associated with this function, and the easier it will become for us over time.

Memorize only what is necessary. Try to use your working memory efficiently: for example, if you're mentally calculating an addition, don't repeatedly recite the new addends to yourself, but instead focus on keeping the partial results in mind. The same applies to subtraction, multiplication, or division: aim to remember only the minimal elements necessary to continue with the calculation. You'll notice that by following this principle alone, you'll make every arithmetic process much quicker and less strenuous.

Learn to know yourself. Many people find it easier to memorize something if they can visualize it. Others if they repeat it mentally or aloud. So try to understand where and how your brain tends to store information more quickly and effectively, and then act accordingly.

Break it down. The brain typically finds it easier to remember many sets made up of a few elements rather than a single very large set. Therefore, especially when dealing with very large numbers, imagine them as being divided into many smaller numbers (or, physically divide them in writing, if you have the chance), and you will have a much easier time keeping them in mind.

Use your hands. When memory fails, a little aid to jot down numbers can always come in handy. But even if we don't have paper or pen to note down figures that we might need later, we can easily remember two numbers between 0 and 9, or one number between 0 and 99, using just our hands.

All you need is the "Thumb Rule," a technique that will be incredibly useful in situations where you need to "set aside" key numbers, such as partial results from addition or carry-over figures in division.

But how does it work? Simply, for each hand, do this:

- To indicate the numbers from 0 to 5, simply keep an equivalent number of fingers raised.
- To indicate 6, place your thumb on your index finger.
- To indicate 7, place your thumb on your middle finger.
- To indicate 8, place your thumb on your ring finger.
- To indicate 9, place your thumb on your pinky.

One hand here will obviously be enough to memorize the first 10 digits, while the other might prove useful if you want to reach up to 100. In that case, you just need to assign tens to one hand and units to the other.

Use phonetic conversion. Phonetic conversion is likely the most well-known and used numerical memorization technique, given its (relative) simplicity, combined with extreme power and versatility. I believe its substantial advantage lies in the fact that, through its use, even a very long number, which is therefore difficult to memorize, can be transformed into a word or phrase that, due to its associable meaning, can generally be much easier to remember.

This technique consists of three phases:

1. You need to associate each digit of the number to be remembered with a consonant (or a set of consonants) through a table that must be fully memorized, which I will show you very shortly.

2. The obtained consonants should be combined with vowels "as you like," in order to create a word or phrase that you will then need to remember well.

3. When you want to recall the number, you will need to bring the memorized phrase to mind, remove the vowels, and use the conversion table you've memorized in reverse.

Here, therefore, is the phonetic conversion table that you need to memorize to master the technique:

0 = S, Z

1 = T, D

2 = N

3 = M

4 = R

5 = L

6 = J, SH, CH, soft G (as in "George")

7 = K, hard C (as in "Cat"), hard G (as in "Go"), Q

8 = F, V

9 = P, B

But since examples are the best way to help internalize a method, let's do one right away: suppose you have to memorize the number 374992. We have:

3 = M

7 = hard C, K, hard G

4 = R

9 = P, B

9 = P, B

2 = N

By adding vowels to these consonants, we could create something like **"McRib Bone,"** which is theoretically easier to remember than 374992. It doesn't matter whether the phrase makes sense or not; in fact, the most bizarre phrases may even be easier to remember. Conversely, let's say we remember that the word associated with our number is "Fireball." Breaking it down:

F = 8

R = 4

B = 9

L = 5

L = 5

By referring back to the table, we can immediately see that the number we need to remember is 84955.

This technique may seem a bit challenging to use at first, but with the right training and practice, it can lead to amazing results. Not to mention, this technique alone might be worth the price of the entire book, since not only will it allow you to perform any calculation without the aid of pen and paper, but it will also be incredibly useful for remembering dates or phone numbers!

Now you have a lot of useful tools at your disposal to improve your "numeric memory." Combine the ones that work best for you and use them to enhance your calculation skills!

3 - Strengthen and Enhance Your Foundations

All our knowledge, including mathematical knowledge, can be seen as a construction made with LEGO: the building blocks of simpler concepts can be assembled to form more complex concepts. These, in turn, can be combined to create even more complex structures. And so on.

It follows naturally that if the initial building blocks are not well-made, the final structures will inevitably be unstable. In other words, if you don't have at least all the basics of arithmetic firmly fixed in your long-term memory, any reasoning that tries to emerge from these basics may be slow, confused, or even completely incorrect. This point is particularly "critical" precisely because it is underestimated in 99% of cases. I recall the example of many of my engineer colleagues, who often failed complex mathematical analysis exams not because they didn't understand the more advanced concepts of the subject, but because they made very simple arithmetic errors.

Therefore, a key to greatly enhancing and accelerating one's calculation abilities is precisely the approach of making "complex structures our new building blocks." This involves making basic reasoning so inherent to our way of thinking that we can quickly and effortlessly construct advanced ones. For instance, if we memorize and train ourselves to instantly recall the results of all additions from 1+1 to 9+9, we won't need to spend a single second on a potential 7+8 in a six-digit addition, allowing us to perform the entire operation much more rapidly.

For this reason, my advice is to work on "cementing" the following concepts in your mind, no matter how obvious or trivial they may seem to you:

The Fundamental Properties of Arithmetic Operations

The fundamental properties of arithmetic operations will be used frequently in a "creative" manner to develop various quick calculation strategies. For this reason, it could be very useful for you to delve into and firmly fix them in your mind. However, if these things are already extremely clear or obvious to you, feel free to completely "skip" them.

Commutative Property of Addition and Multiplication

Changing the order of operations in multiplication or addition does not change the result. So, for example:

$$3 + 4 + 5 = 5 + 4 + 3 = 5 + 3 + 4 = 12$$

Associative property of addition and multiplication

Changing the grouping of numbers in multiplication or addition does not alter the result. So, for example, (3 + 4) + 5, where the position of the parentheses implies calculating 3 + 4 first and then adding 5, is no different from 3 + (4 + 5), which implies calculating 4 + 5 first and then adding 3.

Distributive property of addition with respect to multiplication

If you come across an operation like a x (b + c), the result will be ab + ac

Invariant property of division

Given an a / b, division if you multiply or divide a and b by the same non-zero quantity c, the result does not change. For example:

$$30 / 10 = 3$$

Dividing both numbers by 10, we get

$$3 / 1 = 3$$

Multiplication has a similar property, with the key difference that to keep the result of any "a x b" unchanged, you should no longer divide or multiply both by the same amount "c," but instead multiply one and divide the other, or vice versa. For example, if you have 16 x 2, you can halve 16, but to maintain the result, you need to double 2. Hence:

$$16 \times 2 = 8 \times 4 = 32$$

Existence of the Neutral Element

Adding or subtracting 0 to/from a number leaves the number unchanged. The same applies if you divide or multiply the number by 1. Simple as that, let's move on!

Product cancellation

Any quantity multiplied by 0 always results in 0. Conversely, you cannot get 0 from multiplication unless at least one of the factors is zero.

It's also useful to remember here that the nullification of the product means it is absolutely impossible to divide by zero. Let's consider any number other than 0, for example, 14. Since division is the inverse operation of multiplication, calculating 14 / 0 is like asking: *"What number, multiplied by 0, gives us 14?"* And obviously multiplication by zero always results in zero, making this question unanswerable.

And what if we wanted to calculate 0 / 0? In this case, the answer is said to be "indeterminate," meaning the operation can yield an infinite number of results. In fact, the question: "Which number, multiplied by 0, gives me 0?" has simply infinite possible answers due to the zero-product property. This, mind you, does not equate to an "infinite" quantity, but to the very impossibility of defining the given amount, whether finite or infinite; and yes, if this is already true in the daily matters of life, the idea that in a rigorous field like mathematics there are problems without answers is something that can easily leave some of us with a real "sense of emptiness." Our brain, after all, in order to navigate hostile environments and ensure its survival, has been literally "programmed" by evolution with a genuine "hunger for truth," and tends to detest all incomplete, partial answers, and all absences of them. However, as we will also see when we examine the "historical pill" related to Gödel's Incompleteness Theorems, the impossibility of having all the answers is a "mathematical burden" with which one must necessarily live, even though, quoting a 1989 adventure film masterpiece, "that doesn't mean we have to like it."

Dividing or multiplying by 10, 100, 1000, and other powers of 10

These are also quite simple operations, and for this very reason, the recommendation mentioned a moment ago remains the same: if these concepts seem excessively obvious or trivial to you, always feel free to skip them and move on to the next section.

Multiply by a power of 10

Multiplying a number "a" by a power of 10 (that is, a number made up of "1" followed by only zeros) involves adding as many zeros to "a" as there are zeros after the one. If "a" contains a decimal point, you need to move the decimal point to the right by as many positions as there are zeros after the one. And if, in moving it, the decimal point reaches past the units digit and there are more positions left to shift, you need to "fill" them by adding zeros.

So, for example:

5 x 1000 = 5000 (We added three zeros)

1.28 x 1000 = 1280 (We moved the decimal point two positions to the right to reach the units place. The zeros in 1000 are three, so there's *one* position left to move, and hence we will put *one* last trailing zero.)

Divide by a power of 10

To divide a number "a" by a power of 10, you could do the following:

1. If the number does not have a decimal point, to simplify, you could imagine it as having a "point zero." So, for example, 377 would become 377.0.
2. Move the decimal point to the left by as many positions as there are zeros in the divisor.
3. If the decimal point ends up to the left of the original number, add a zero before the decimal point; if you haven't exhausted the positions to move, repeat this rule.
4. At the end of the process, remove any zeros after the decimal point, and if there are ONLY zeros after the decimal point, remove the point.

So, for example:

345 / 1000 = 0.345 (moving three places to the left, the decimal point jumps over the 3. Therefore, a "zero" is placed in front)

2 / 1000 = 0.002 (moving three places to the left. First we pass over the 2, resulting in .2 = 0.2. Then, we continue moving left, passing over this zero, giving us .02 = 0.02. Finally, at the third position we move again, resulting in 0.002)

Being able to quickly perform multiplications or divisions by powers of 10 also helps to swiftly manage multiplications or divisions by any number that ends with zeros. Specifically, in the case of multiplication where one of the factors ends with one or more zeros, you can remove the zeros from that number, perform the operation with the "truncated" number, and then multiply the result by the power of 10 corresponding to the number of zeros you had removed.

So, for example, if we want to multiply 350 x 2000, we can remove all the zeros and calculate 35 x 2 instead. In this case, I removed four zeros in total, so I must multiply the final result by 10,000 to restore the original value.

In the case of division between numbers that end with zeros, we can operate in a similar way, but with an important distinction:

- All zeros removed from the dividend (the number before the division sign) mean that, to reach the correct result, you must multiply the final number by the power of 10 corresponding to those zeros.
- All the zeros removed from the divisor mean that the final result must be divided by the power of 10 corresponding to those zeros.

And so, for example:

90 / 6 = We can remove the "0" from 90 and calculate 9 / 6. The result is 1.5. Since I removed the zero from the dividend, I must multiply 1.5 by 10, obtaining the final result of **90 / 6 = 15**.

9 / 60 = We remove 0 from the divisor and then have to divide by 10 the 1.5 resulting from **9 / 6 = 0.15**

Addition and multiplication tables from 0 to 9

Most of us know the tables that follow, at least in broad terms, but I bet few can recall them instantly and without mistakes. So, try to review them if necessary, paying special attention to memorizing all the "points" that you find more challenging. For example, it is common knowledge that most people tend to encounter problems with the multiplication tables of 7 and 9, as well as make mistakes with single-digit additions involving numbers from 6 to 9. Let's also clarify, in case it's not obvious, that due to the commutative property of addition and multiplication, only the lower half of the tables that follow has been filled in, since the upper half would be identical.

Addition table

+	0	1	2	3	4	5	6	7	8	9
0	0									
1	1	2								
2	2	3	4							
3	3	4	5	6						
4	4	5	6	7	8					
5	5	6	7	8	9	10				
6	6	7	8	9	10	11	12			
7	7	8	9	10	11	12	13	14		
8	8	9	10	11	12	13	14	15	16	
9	9	10	11	12	13	14	15	16	17	18

Multiplication table

X	0	1	2	3	4	5	6	7	8	9
0	0									
1	0	1								
2	0	2	4							
3	0	3	6	9						
4	0	4	8	12	16					
5	0	5	10	15	20	25				
6	0	6	12	18	24	30	36			
7	0	7	14	21	28	35	42	49		
8	0	8	16	24	32	40	48	56	64	
9	0	9	18	27	36	45	54	63	72	81

The "three fundamental steps" to improve your calculation skills end here. By maintaining the right attitude, training your memory, and strengthening your foundations, you will enhance your "mathematical" abilities and more, in ways you probably never even imagined. And the beauty of it all is that this is just the beginning of your journey.

The Historical Pill

In Mesopotamia, the ancient Sumerians developed, around the third millennium BC, a more advanced numerical system than the Egyptian one, based on 60 (the so-called sexagesimal system). This choice is likely due to the fact that 60 has many "useful" divisors for practical applications (consider the ability to easily divide by 2, 3, 4, 5, 6, 10, etc.), although there is a theory that this might derive from the somewhat challenging possibility of counting to 60 using fingers, considering the front and back phalanges. Their numerical writing system could also seem terribly inaccurate to those not specifically trained in it; for example, a vertical wedge could mean 1, 60, 3600, or 216,000, depending on the context. The "incredible" thing, if you will, is that the sexagesimal system is one we still use today, after so many millennia: consider the "base 60" of time or angles, for example.

The Aphorism

"In mathematics, patterns can be found everywhere. It's the universe. I cannot imagine anything without numbers. It's beauty everywhere. It's a sort of paradise."
(Maryam Mirzakhani)

Mathematical Oddities

Various animals have demonstrated an understanding of the concepts of quantity and size. Bees, for example, seem to be able to communicate to each other information about the distance between the hive and a food source, such as a pollen deposit.

Some species of birds are also able to discriminate between different sets of objects or respond to signals that indicate different quantities. Some primates, such as chimpanzees and bonobos, have been shown to understand numerical order and even solve simple math problems. And if they can do it...

IV - Instantly Multiply with Your Fingers

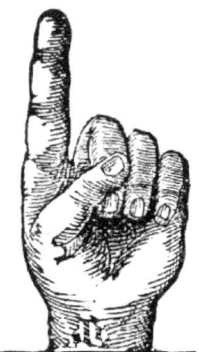

Everyone knows that it is possible to perform small additions with one's hands. After all, it's one of the very first things we learn at school. But did you know that with your hands you can also do some... simple multiplications?

If you suddenly experience a mental blackout during a complex operation and can't remember a particular multiplication table, you can use two amazing hand tricks that will allow you to instantly perform a wide range of single-digit multiplications.

Let's start with the first very simple technique, which takes advantage of the fact that the digits of any multiple of 9 up to 90, when added together, equal 9. Specifically, this technique will allow you to immediately perform all multiplications from 9 x 1 to 9 x 10. Here's how it works and how it can be applied:

- Turn both your palms towards you, pointing your fingertips upwards.

- Each finger is simply assigned a number corresponding to their order with the palms facing you. Thus, the left thumb will be 1, the left index finger 2, and so on, up to the right thumb which will be 10.

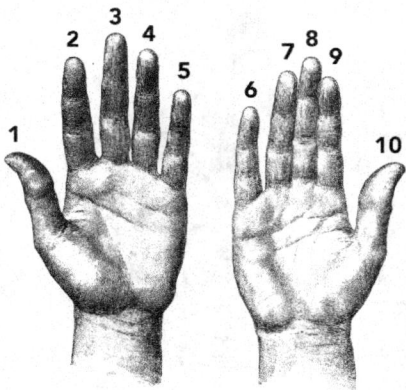

- To calculate 9 times "a," lower the "a" finger.
- The first digit of the result will be equal to the number of fingers before the one lowered. The second digit of the result will be equal to the number of fingers after the one lowered.

So if, for example, we wanted to calculate 9 x 1, we could simply lower the left thumb. There are no fingers before the thumb, and therefore no tens. However, after the left thumb, there are 9 fingers, which correspond to the units of the result. And, as most of you may know, 9 x 1 = 9.

If instead we wanted to calculate 9 x 6, we would lower the right pinky finger. Before the right pinky finger, there are five fingers, so the first digit is 5. After the right pinky finger, there are 4. And indeed, 9 x 6 = 54.

As you can see, it's an extremely basic strategy, which can also be remarkably effective for all those children who struggle to remember the 9 times table at school.

Now, let's move on to the second technique, which is also very easy to learn and use, and allows us to instantly calculate all the times tables from 6 x 6 to 10 x 10. Here's how to do it:

- Always turn both palms of your hands towards you, but no longer point your fingers upwards; instead, rotate your left hand to the right and vice versa, so that the fingertips of each hand face the other hand.
- For both hands, mentally assign 6 to the pinky, 7 to the ring finger, 8 to the middle finger, 9 to the index finger, 10 to the thumb.

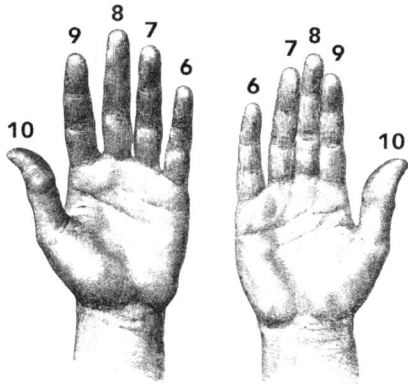

- If you want to multiply one number by another, make the corresponding fingers touch each other, being careful to maintain the palm orientation I described. So:
- 7 x 8 = ring finger touches the middle finger.
- 10 x 6 = thumb touches the little finger, and so on.
- Now, keep them like that and count: the fingers below the "contact" point, where the fingers touched, plus those belonging to the contact point itself represent the "Below" section. Count the fingers in the "Below" section. They must be at least 2.

- The "free" fingers above the contact point represent the "Above" section. Now, count separately the fingers in the "Above" section of one hand, and those on the other.
- To recap for a moment, you should now have in mind three "partial results": count "Below," count "Above" right, and count "Above" left.
- Multiply the "Below" count by 10. Keep the result in mind.
- Multiply "Over Rights" by "Over Lefts." Keep this second result in mind.
- Now, add the two results together.

Since this technique is slightly more complex than the previous one, let's proceed with some examples. Let's say we want to multiply 6 by 7. We will proceed as follows:

- Let's touch the left pinky with the right ring finger (or vice versa).
- "Below" is equivalent to 3 fingers (the 2 at the point of contact plus the right pinky). Multiplied by 10 equals 30.
- "Above rights" is equal to 3, while "Above lefts" is equal to 4. Multiplied together, they equal 12.

$$30 + 12 = 6 \times 7 = 42$$

Second example, 8 x 7:

- Let's touch the right ring finger and the left middle finger (or vice versa).
- "Below" is equal to 5 fingers. Multiplied by 10 equals 50.
- "Above right" equals 2. "Above left," on the other hand, equals 3. When multiplied, they give 6.

$$50 + 6 = 8 \times 7 = 56$$

As mentioned earlier, this second technique can also be used for the 9 times table, but obviously, in that case, it is better to use the "far more efficient" first strategy. Additionally, it is clear that what has been discussed so far cannot be utilized if one is not able to quickly and instantly perform multiplications of numbers up to 6. This shouldn't be a major problem, however, since, as suggested in the previous chapter, most of us tend to encounter the greatest difficulties precisely with multiplications between numbers approaching 10.

With your fingers, you can also apply two very small mathematical tricks that could be potentially useful in everyday life:

- **Measure any object of any length.** It's simple, yet an extremely useful little "life hack": measure one of your fingers with a ruler, remember the result, and the next time you need to measure any object without a tape measure on hand, just count how many times your finger "fits" into the object, then multiply that number by the length of your finger.

- **Measure the height of a tree, a building, or any other large object:** no, if you need to measure a tree or a building, you don't have to see how many times your finger "fits" in it. Simply use this technique that's been around since ancient times: place a person whose height you know next to the building. Position your finger close to your eye such that its height matches the person's visual size, and then apply the same trick as before! Are the calculations getting tricky here? Don't worry, the techniques in the upcoming chapters will soon come to your aid!

Now that you have mastered the basic foundations for entering the world of rapid mental calculation and have already begun to see some small practical applications, we will introduce a fascinating topic that is highly cherished by both mathematics enthusiasts and those interested in Asian culture: Vedic Mathematics.

The Historical Pill

The ancient Egyptians were among the first to develop, around 2700 BC, a decimal numbering system, which, however, was based on hieroglyphics and therefore was not positional like ours (i.e., where the value of digits changes depending on their position). The choice of the base being 10 is very likely linked to the number of fingers on the human hands, and to the ease with which it is possible to count and perform simple calculations using them. This, also, almost certainly explains why base-ten numbering systems are the ones that have "better survived" through the ages, becoming the main standard in our era.

The Aphorism

"Mathematics is a magnificent and vast landscape open to all who find joy in thinking, but not well-suited for those who do not love to think."
(Immanuel Lazarus Fuchs)

Mathematical Oddities

Prime numbers, being divisible only by 1 and themselves, not only have extraordinarily interesting applications, but have also often been literally "adopted" by nature to optimize certain processes. Some cicadas, for instance, emerge only after prime-numbered years to avoid predators with life cycles that are divisors of their emergence periods evolving to specialize in capturing them.

In our world, prime numbers are fundamental, among other things, for cryptography. In fact, the security of many of our digital transactions, such as online banking operations or sending encrypted emails, relies on the use of very large prime numbers. This is because the "prime factorization problem" (i.e., the task of breaking a number down into its prime factors) becomes incredibly difficult and requires an enormous amount of computational time when dealing with very large numbers. This, therefore, makes it almost impossible for anyone to try to decrypt the encrypted information without knowing the original key.

V - Mathematics and Hindu Wisdom

In the 1960s, the book *Vedic Mathematics* was published posthumously, attributed to the Indian monk and scholar Bharati Krishna Tirtha. This manual, which presents a series of techniques to speed up arithmetic calculations, was promoted as a revelation of ancient Vedic mathematical knowledge, although, to be honest, none of these techniques appear to be historically connected to the Vedic period. In fact, the very title of the book, chosen by publisher V. S. Agrawala, has drawn criticism for its inaccuracy, as neither Tirtha nor Agrawala could ever prove the Vedic origins of the techniques presented.

Nonetheless, the Vedic approach to certain mathematical problems has been widely used and appreciated for quite some time now. Over recent decades, it has spread extensively, revealing a significant added value and captivating people with its ability to encourage, among other things, individual expression and creativity. Many of the mathematical strategies that follow in this chapter, in fact, will draw directly from the Vedic Sutras, often combining them to create quick, creative, and highly effective calculation methods.

But let's begin by explaining the meaning and use of the so-called "Sutra Nikhilam," which states: *"All from 9 and the last from 10."*

This Sutra reveals a very quick and effective method to calculate the distance (or difference) between any given number "a" and the next higher power of 10. For example, between 1000 and 450, between 100 and 67, between 10,000 and 9876. Its usefulness, as we will see shortly, is not limited to this, but let's proceed step by step.

To apply the Sutra in its "essential" form, you can proceed as follows:

- Start with a subtraction where the subtrahend (the number following the minus sign) has as many digits as there are zeros in the power of 10 from which you want to subtract. If this is the case, you can use the Sutra. For example, you can use it to calculate 1000 - 345, or 10,000 - 6777.

- Now, take only the subtrahend (in the examples just seen, 345 or 6777) and starting from the left, subtract each digit of it from 9 (so perform, careful, "9 minus the digit" and not the opposite), except for the units digit, which you will subtract from 10. Then, place the results of these subtractions into the result.

- Whenever you end up with a "10" as a result of this subtraction (which happens, for example, when the units digit is zero), you should add a 0 to that digit and simultaneously add a 1 to the leftmost digit.

To better clarify what we've seen, let's go with an example and assume we want to calculate 10,000 - 4,350. We start with the first digit on the left of 4,350, and we will have:

- Subtract 4 from 9 to get 5.

- Subtract 3 from 9 to get 6.

- Subtract 5 from 9 to get 4.

- Subtract 0 from 10 to get 10 (write down 0, carry over 1 to the previous result).

- This results in 5,650.

With the traditional methodology we learned in our schools, a subtraction like this could have required a much longer and more complex process. Yet, with these few simple steps, it is possible to greatly speed up the process, minimizing the number of necessary steps and, consequently, the likelihood of making errors.

Now let's extend the Sutra and use it to perform other types of subtractions. For example, given any subtraction a - b, to derive the result:

- Use the Sutra on B.
- Determine the power of 10 that is immediately greater than "b."
- If "a" exceeds that power by a certain amount, add that amount to the final result of the Sutra.
- If, on the other hand, "a" is less than the aforementioned power of 10 by a certain amount, subtract that amount from the result.

This technique will be all the more useful where the difference in "a" and the power of 10 in question is quick to calculate or intuit. But since all this can be much easier done than said, let's solidify this concept with some brief practical examples:

- To calculate 1002 - 456, for example, we can use the Sutra on 456 (=544) and add 2 to the result (1000 is the power of 10 immediately greater than 456 and 1002 exceeds 1000 by 2. So, 544 + 2 = 546).
- Alternatively, to calculate 20,004 - 3,498, we can use the Sutra on 3,498 (which equals 6,502) and add 10,004 (10,000 is the power of 10 immediately higher than 3,498, and 20,004 exceeds 10,000 by 10,004. 6,502 + 10,004 = 16,506).
- Or again, to calculate 9998 - 4432, we can quickly use the Sutra on 4432 (= 5568) and subtract 2 from the result (10,000 is the power of 10 immediately higher than 4432, and 9998 is 2 less than 10,000. 5568 - 2 = 5566)

A practical application, which I find very useful, that comes to mind for these strategies is quickly quantifying the expenditure made from an initial budget, which is very often a "round" number, making it well-suited to similar calculation methods.

For example, if before starting a project we had 7000 dollars in my account, and now, after a month, we will have 1288 only, we can first use the Sutra and then subtract 3000 to quickly quantify the expenses incurred. We will find out that the monthly expense was 8712 - 3000 = 5712 dollars.

Alternatively, let's try to apply this Sutra in a case related to personal financial analysis. Suppose you have a certain initial savings amount, say 10,000 dollars, and you decide to invest part of it in stocks or other financial instruments. After a certain period of time, you notice that the value of your investment has changed and, unfortunately, has decreased to 8,743 dollars. By applying the Nikhilam Sutra, we get 1,257 (10,000 - 8,743). To calculate the percentage loss relative to the initial investment, simply divide the loss by the initial investment and multiply by 100. So we would have:

$$(1{,}257 / 10{,}000) * 100 = 12.57\%$$

This leads us, sadly, to acknowledge our loss and (probably) reconsider our portfolio allocation strategy.

But the applications of Vedic Mathematics don't end here, as we will see later; it will prove very useful when we begin to discuss rapid multiplication.

The Exercise

If you feel like practicing applying this "Sutra" to perfection, try using it to calculate:

10,000 - 4,351 = ?

1,000 - 673 = ?

1,000,000 - 456,789 = ?

200,000 - 12,285 = ?

3,000 - 99 = ?
40,000 - 1,111 = ?
900,000 - 98,765 = ?

The Historical Pill

The history of ancient Chinese mathematics is fascinating, filled with innovations and discoveries that have had a lasting impact on the development of the discipline worldwide. One of the most famous books is the "Zhou Bi Suan Jing" or "The Book of Numbers and Measurements," one of China's oldest mathematical works. It is thought that its composition occurred in several stages starting from 1046 BC, but its final form may date back to a period between 300 and 200 BC. This fundamental text includes a series of mathematical problems along with their solutions and contains a theorem similar to what is known in the West as the Pythagorean Theorem. However, ongoing historical debate suggests that this theorem was not known in China as early as 1046 BC, but rather is part of later additions and commentary.

The Aphorism

"Mathematics is the most economical of the sciences. Unlike physics or chemistry, it requires no expensive equipment. All you need for mathematics is a piece of paper and a pencil."
(George Pólya)

Mathematical Oddities

The problem of finding a formula to "quickly discover" prime numbers is one of the oldest and most persistent riddles in mathematics. There is still no known formula or highly efficient algorithm that allows one to extract, generate, or identify prime numbers, which seems to be due to the fact that prime numbers appear to be distributed almost "randomly" along the line of natural numbers, making it extremely difficult to predict the next prime number in a sequence. Despite this, there are various techniques, although inefficient, that can help identify prime numbers within a certain range, such as the sieve of Eratosthenes. This technique involves creating a list of numbers from 2 up to the maximum number you wish to analyze; you then take the first number on the list (in this case 2), declare it a prime number, and strike out all of its multiples from the list. You take the next number, do the same, and so on. At the end of the process, all the numbers that remain on the list will "necessarily" be prime numbers.

VI - Composing, Decomposing, and BlackJack

This chapter will teach you three essential techniques to perform additions and subtractions more rapidly, and then present you with a particularly interesting application. You will very likely realize, as you read through these techniques, that you have already used them at times in the past without even noticing. However, becoming fully aware of them will help you better understand how and when to use them effectively.

Searching for tens and squares

Let's begin by immediately explaining the first technique, which is very simple and useful, especially when you have to add (or subtract) many small numbers together: group the addends or subtrahends in such a way that they form multiples of 10.

For instance, let's say you need to calculate

$$3 + 4 + 5 + 6 + 7 + 2 + 5 + 1 + 3 + 4 + 3$$

The first thing you can do is combine the first three and the fifth seven, hence obtaining:

$$10 + 4 + 5 + 6 + 2 + 5 + 1 + 3 + 4 + 3$$

Now we can take the second-to-last four, the last three, and the third-to-last three. Added together, they equal ten again, and so we have:

$$20 + 4 + 5 + 6 + 2 + 5 + 1$$

Once again, by adding four, five, and one, we get a third 10, and therefore we've got:

$$30 + 6 + 2 + 5 = 43$$

Beyond the obvious practical simplicity, this method has the advantage of being an essentially "creative" approach, especially for children who are just beginning to learn the subject: searching for the numbers to make a ten can also be fun and help young ones become familiar with the method.

If we then wanted to add a little "creative challenge" to this method as well as to any other additive sequence of more or less simple quantities, we could exploit the principle according to which *the sum of N consecutive odd numbers, starting from 1, is simply N^2.*

Returning to our initial addition (which was 3 + 4 + 5 + 6 + 7 + 2 + 5 + 1 + 3 + 4 + 3) we can take the 1, a 3, a 5, the 7, and "create" a 9 by combining the 6 and another 3. From which:

$$1 + 3 + 5 + 7 + 9 =$$

(these are the first 5 odd numbers and hence it equals 5^2)

= 25 + (combining the other numbers) 18 = 43

Borrowing technique

This second technique is quite similar to the previous one; it can be applied to additions and subtractions of any kind. It involves "borrowing" a quantity from one of the addends (or subtrahends) and adding it to another addend (or subtrahend) in such a way that one of the two operands becomes a power or multiple of 10, thereby theoretically simplifying the final calculation.

Let's say, for example, we have 368 + 214. One thing you can do immediately to simplify the calculations a bit is to borrow a "2" from the second addend, from that 214, add it to the first, and get 370 + 212, which can be calculated almost at a glance.

Same situation if we have 967 - 255. We could think of "borrowing" 3 from the 255, getting 252. We give that 3 to 967, which becomes 970, giving us a chance to calculate an equivalent subtraction 970 - 252 = 712 much more quickly? *No,* be careful here because with subtraction the rules change subtly: in the previous example, we took an amount from the number after the sign, so no additional operations were necessary. However, if we need to do the opposite and have the borrowed amount "navigate" in the other direction, any amount taken from the subtrahend and passed to the minuend must change its sign. So, going back to our good old 970 - 258, a possible 3 "borrowed" from 258, passed to 970 and having its sign changed, will result in 967, not 973. Hence,

970 - 258 = 967 - 255

(and not 973 - 255)

Another note: in order to calculate that 970 - 258, you could even use the Sutra I mentioned in the previous chapter, and this is a clear sign that the real strength of the tools contained in this book lies in how they can be freely recombined to create calculation strategies suitable for every kind of situation.

- The complement of 970 to 1000 is 30. Let's keep this number in mind, as we will need at the last step.
- Applying the Sutra to 258 will produce 742.
- Once again, 742 - 30 = 712.

The "borrowing rule" also allows us to establish these basic simplification guidelines that, as common sense as they may seem, are always better to "refresh" so they can be quickly recalled from memory when needed:

- **To add 9 to a number, add 10 and subtract 1.**
- **To add 8, add 10 and subtract 2.**
- **To add 7, add 10 and subtract 3.**
- **To subtract 7, subtract 10 and add 3.**
- **To subtract 8, subtract 10 and add 2.**
- **To subtract 9, subtract 10 and add 1.**

And the reasoning to apply will clearly be identical for numbers like 90, 80, 900, 800, etc.

Break that down

The last technique, which is also very simple and particularly useful when dealing with operands of at least three digits, involves breaking down one or more of the operands and working separately on its units, tens, hundreds, and so on.

That is, quite simply, if you need to calculate 3456 - 1734, it will be much easier to work by breaking down 1734, and then subtracting first 1000, then 700, then 30, and finally 4.

The human mind, in fact, is capable of performing many simple operations much more quickly and accurately than a single complex operation. It is precisely on this principle that many other rapid calculation strategies we'll see in the future will be based.

As another example, if we need to calculate 482 + 389, we can proceed by breaking down 389, that is:

$$482 + 300 = 782$$
$$782 + 80 = 862$$
$$862 + 9 = 871$$

But now that we have introduced the three fundamental strategies for quick addition and subtraction calculations, let's talk about a well-known and entertaining application of these techniques: card counting in Blackjack.

If you're not aware of what this is about, let's start with a brief introduction: all gambling games are designed in such a way that, in the long run, it will always be us losing and the casino pocketing our losses. This can be said as: "Every gambling game has a *negative expected value*," as we will see in the chapter dedicated to probability theory.

A rare exception to this reality is Blackjack, where 'counting cards' can lead to a positive expected value for the player, potentially resulting in increasing winnings over time.

Blackjack, by the way, is a very simple game: for those who are not familiar with it, it is played with French cards, and both the dealer and all players are dealt a face-down card. Face cards are worth 10, aces can be worth 1 or 11 (at the player's "discretion"), and all other cards retain their face value.

Each player, on their turn, can choose to "stay" with what they have or take another card from the dealer, aiming to get as close as possible to 21, but never exceeding it. In fact, anyone who goes over 21 is said to "bust" and loses their bet, which immediately goes to the dealer.

When each player who hasn't busted has determined their score, it's the dealer's turn to make a move. The dealer must draw cards until their total is at least 17, and they must stop if the total reaches 17 or higher. If the dealer busts, they must pay the bet to every player still in the game. Otherwise, they must pay the bet to those with a higher score than theirs and collect it from those with a lower score.

Now that I'm (more or less) sure you know what I'm talking about, let's list some good news about card counting:

- Counting cards is simple. Or rather, we will introduce a type of counting that is very simple, called "High/Low."
- Counting cards is not illegal.
- If you've seen any Hollywood movies where someone caught counting cards gets beaten by the local mob, rest assured and keep in mind that it's all cinematic fiction.

Now, the bad news:

- Even though it's unlikely they'll step on your toes, if they catch you doing it, you might be asked to leave.
- In virtual casinos, card counting is impossible since the decks are shuffled virtually.
- In physical casinos, every effort is made to ensure that if the expected value of card counting is positive, it is kept very low.

Having said that, why should we be interested in this activity?

- Because honing your calculation skills with card games is much more fun than doing it in front of books.
- Because if you keep a low profile and don't abuse the method, in some small casinos it's still possible to cash in a little money without risk and without too much effort.

Let's begin then. Let's say we sit at the Blackjack table. We'll start our count from 0, and as we see new cards being dealt on the table, we'll proceed as follows:

- **When we see 2, 3, 4, 5, 6, we add 1 to our count.**

- **When we see 10, Jack, Queen, King, Ace, we subtract 1 from our count.**
- **When we see 7, 8, 9, we leave the count unchanged.**

This is a simple calculation that must be done quickly and benefits from the initial two calculation strategies introduced earlier.

As we keep track of the count, we must consider that a high total (+6/+9) indicates a deck loaded with high cards and face cards, which is advantageous for us and suggests we can bet high. On the other hand, a low total indicates that we are more likely to lose against the dealer.

In particular, here is an example of an operational strategy that is both cautious and effective:

- **Count equals 0 or is negative:** do not bet.
- **Count equals +1:** go with your base bet (any "minimum bet" previously chosen as such.)
- **Count equals +2 or +3:** double your base bet.
- **Count equals +4 or +5:** triple your base bet.
- **Count equals +6 or +7:** quadruple your base bet.
- **Count equals +8, +9 or higher:** place five times your base bet.

The Exercise

Take a deck of playing cards, draw the cards one by one and practice maintaining the "high/low" count as you draw them. Stop at the twentieth card, shuffle, try again and train yourself to do it increasingly faster.

The Historical Pill

The Plimpton 322 clay tablet is a fragment of a Babylonian clay tablet dating back to around 1800 BC, long before the advent of Greek mathematics. It contains a list of "Pythagorean triples," which are groups of three numbers that satisfy the Pythagorean theorem (the square of the longest side of a right triangle is equal to the sum of the squares of the other two sides). This is a more significant, although still debated, instance of the foundational principles of the Pythagorean theorem known (even if not proven or formally stated) long before Pythagoras himself.

The Aphorism

"If people do not believe that mathematics is simple, it is only because they do not realize how complex life is."
(John von Neumann)

Mathematical Oddities

Although "1" fits the definition of a number divisible only by 1 and itself, it is never considered a prime number because doing so would undermine the foundations of various mathematical theorems. For instance, if we regarded "1" as a prime number, the Fundamental Theorem of Arithmetic, which states that every number greater than 1 is the product of prime numbers in only one way (apart from the order), would lose its truth and consistency. This is because "1" is the result of an infinite number of possible multiplications between prime numbers, starting with the multiplication of two "1"s up to involving an infinite number of them. Consequently, this would likely force us to completely reevaluate numerous postulates and practical applications, such as those in the field of cryptography.

VII - The Magic Column

Let's reiterate a fundamental point about calculation methods: a calculation method greatly benefits in terms of speed and efficiency when it is designed in such a way that it forces the brain to memorize *as few pieces of information as possible*.

Let's imagine, for example, having to mentally calculate the following operation with a ":

$$\begin{array}{r} 989\ + \\ 724\ + \\ 102\ + \\ 670\ + \\ 112\ = \end{array}$$

If we were to perform column addition using the traditional method learned in school, we would first need to add the units separately, followed by the tens and hundreds, while keeping track of several carry-over numbers essential for the final result. This process, particularly if we do not have the option to jot down partial results on a piece of paper, can turn out to be a highly complex task.

Precisely for this reason, when calculating column additions entirely in your head, you can use the "magic column" technique, which allows you to minimize the numbers you need to remember. Here's how it works:

- Instead of starting from the units, as you would for a normal column addition, start at the top left, move down, and gradually add all the numbers in the same left column together, making sure to keep track of only the partial results of the addition each time. In the case of the column addition illustrated just above, as you descend the column you'll find: "9 (+ 7 =), 16 (+ 1 =), 17, 23, 24" (and here you can easily perform the partial operations using some of the techniques illustrated in the previous chapter).

- You have completed the operations on the leftmost column and your partial result is 24. Now move one column to the right (in this case, the middle column), take the first digit, place it to the right of the previous partial result (so you would have 24_8) and, while temporarily leaving the old partial result intact, add each of the digits below it in the same column to this figure. So, using the same example, repeat in your mind: "24_8, 24_10, 24_10, 24_17, 24_18." Only if, as in this case, you end up with a number greater than 10 at the end of the addition, place only the units in the new partial result and add the tens to the leftmost digit. Therefore, your new partial result is 24_18 = 258, after carrying the 1 from 18 to the nearby 24. We are close to the final result.

- Take once again the partial result from the old column and add the top number from the rightmost column (in this case, you have 258_9). Then behave exactly as you did with the previous column: leave the old partial result intact, sum the digits of that column together, and add the carryover only at the end. In this case, you will have: "258_9, 258_13, 258_15, 258_17 = 2597." There are no more columns to the right, so this is the final result.

We have thus achieved the final result in an extraordinarily swift and effective manner, needing to keep in mind just one number at a time and without paying any attention to partial carryovers.

Clearly, if we were to align numbers composed of a different number of digits, we should:

- Align the units with the units, the tens with the tens, etc.
- Act as if the empty spaces, inevitably present to the left of smaller numbers, were filled with zeros. To make things simpler, we could even add zeros to the left of the smaller numbers, so that we end up with numbers consisting of the same number of digits.

Let's now look at another example to better establish the method. Suppose we need to calculate 1341 + 450 + 2451 + 888 + 9872. First, we align the numbers in columns, and for simplicity, we add a "0" to the left of the numbers consisting of three digits:

```
1341 +
0450 +
2451 +
0888 +
9872 =
```

- Let's start from the left and we have: "1, 3, 12." The partial result is therefore 12.
- Moving to the right we have: "12_3, 12_7, 12_11, 12_19, 12_27." The partial result, therefore, is 147.
- Moving further to the right, we have: "147_4, 147_9, 147_14, 147_22, 147_29." The new partial result is therefore 1499.
- Operating on the last column, we have: "1499_1, 1499_2, 1499_10, 1499_12." Repeating the '1' twice, the final result will be 15,002.

Show your friends that you can solve vertical addition problems entirely in your head, but without revealing your strategy! I'm sure that, if you practice doing partial sums quickly and without errors, the "show" effect will be highly impressive!

If you want to move to the "next level" and further improve your calculation performance, you can do two things:

- Practice working with numbers without necessarily arranging them in columns.
- Work even more quickly by operating on multiple columns simultaneously.

Regarding the second point, let's use a practical example to explain what we are talking about. Let's assume we have:

$$
\begin{aligned}
561\ &+ \\
343\ &+ \\
912\ &+ \\
134\ &+ \\
451\ &=
\end{aligned}
$$

We already know how to operate column by column, but if we wanted to speed up the calculations even further, we could:

- Always start from the left, but this time simultaneously add the two-digit numbers from the first two columns, and then begin with "56, (+ 34 =) 90, (+ 91 =) 181, 194, 239"
- Regularly calculate on the last column with "239_1, 239_4, 239_6, 239_10, 239_11." The final result is indeed 2401.

This is clearly the next step, which you will be able to achieve only after you have sufficiently trained your mental flexibility with two-digit additions (thanks also to the strategies from the previous chapter) and have first mastered the technique of the single column.

The Exercise

Try to perform the following column additions completely in your mind:

176 +
384 +
592 = ?

438 +
291 +
127 +
954 = ?

125 +
397 +
582 +
614 +
328 = ?

143 +
285 +
392 +
614 +
582 +
731 = ?

1238 +
4591 +
3827 +
5264 = ?

The Historical Pill

Within the realm of mathematical history, one cannot fail to mention the Pythagorean School, founded by Pythagoras in the 6th century BC, which introduced the idea that numbers and geometry could be used to explain the world in rational and logical terms. The Pythagoreans believed that "everything is number," emphasizing the belief that the world could be understood through mathematics; considering how physics and the mathematical rules at its foundation have allowed us to deeply understand the laws of the universe, this is an extraordinarily contemporary philosophical concept, able to withstand the passage of millennia and arrive substantially intact to the present day. Additionally, we owe to the Pythagoreans the foundations of the "discovery" that harmonious sounds correspond to simple ratios between the lengths of vibrating strings, thus laying a significant part of the groundwork for the creation of Western music as we know it.

The Aphorism

"Mathematics is like the game of checkers, suitable for young people, not too difficult, enjoyable, and without any danger to the state."
(Plato)

Mathematical Oddities

The Pythagoreans believed that the number 1 was neither even nor odd, but both, as it was the generator of all numbers, whether even or odd (it was enough to add it to an even number to generate an odd one, and vice versa). Therefore, it was considered a sort of ideal "container" for both concepts.

VIII - Introduction to Game Theory

Now that we have learned (almost) everything necessary about quick addition and subtraction, let's begin to discuss an extremely useful, interesting, and fascinating practical application: game theory.

Game theory is a branch of mathematics that studies the evolution of situations where there are participants making decisions. It is an incredibly broad field, and thus a comprehensive discussion is beyond the scope of this text. However, we can still provide some insights that will certainly be very useful in the context of decision theory, especially in cases where other people might make certain choices or not, and these choices could significantly influence something of interest to us.

In particular, we will start by mentioning those games where there are two decision-makers (you and another person, for example) and these two can make decisions within a range of two possible choices. Let's therefore begin by analyzing a variant of the "classic" example of the "Prisoner's Dilemma":

There are two men, A and B, who have been investigated for a major crime and taken prisoner. Neither has knowledge of what the other will do. However, they have TWO possible choices: deny or confess, and depending on what they do, the police will act in a specific way. Namely:

- If both deny committing the crime, they will each be given only one year in prison.
- If one confesses and the other denies it, the one who denies will get ten years in prison, while the one who confesses will go free.
- If both confess, they will each be sentenced to five years in prison.

Let's now assume that we are person "A" and understand how this situation can be outlined in a table from "our" point of view.

	Other one confesses	Other one denies
We confess	-1	-10
We deny	0	-5

In this table, the rows correspond to our potential decisions, the columns to the potential decisions of the other person, and the cell where a specific row and a specific column meet corresponds to the years of imprisonment we will receive if those specific choices are made (here the number is indicated as a negative, because years of imprisonment are obviously considered a "disadvantage").

But what are addition and subtraction useful for in such a context? Given a tabular representation of certain situations, it is precisely through these operations that we can choose the strategy to try to maximize our advantage.

The first thing to do in these contexts, in fact, is to see if it's appropriate to adopt a "pure" dominant strategy. In other words, to determine if it makes sense to always make the same choice because it is "universally" more advantageous than the other, regardless of the action taken by the other party.

In particular, we will say that it is advisable to adopt the pure strategy of always making the same choice when every value in the row corresponding to that choice is higher than the corresponding value in the other row.

In this case, for example, a rational approach would suggest that it is always beneficial to confess, given that 0 years of imprisonment is preferable to 1 year, and 5 years is preferable to 10 years. Not coincidentally, we have presented the "prisoner's dilemma" with these specific values to show how, certain "games," certain life situations marked by uncertainty about what may happen "on the other side," can nevertheless be managed by seeking the "pure dominant strategy," a set of actions that will work regardless of the extent of this uncertainty; which can be very important, considering how such gaps in our future vision often risk throwing us into a true decision-making paralysis.

Other times (and perhaps more frequently), the same problem is presented with slightly different values, in which confessing is truly advantageous only if the other person does so as well, and it becomes extremely disadvantageous for both if not; the goal is to model and reflect on situations where potential levels of trust, communication, and cooperation are key elements for optimally resolving an issue. Among other things, this latter case has been studied in biology for analyzing evolutionarily successful behaviors and, in particular, for searching for a possible explanation of how altruistic and cooperative behavior could have emerged in animal species (akin to the idea of "confessing," equivalent to adopting a behavior based on the trust that the other will cooperate), even given the apparent advantage of opportunistic behavior. These opportunists, by exploiting resources and contributions of others without ever reciprocating, would theoretically seem able to thrive and thus appear "far more successful" in an evolutionary context. Richard Dawkins, particularly, in his famous book "The Selfish Gene," discussed the results of studies analyzing the impact of various possible strategies, given similar dilemmas, in repeated contexts. The extremely fascinating concept that emerges from his reflection is not only that all completely "negative" strategies, being "exploitative" and non-cooperative, were gradually "weeded out" throughout the evolutionary process, but also the emergence of a dominant strategy called "Tit for Tat." What is interesting about "Tit for Tat" is its being an extremely simple strategy that outperformed much more complex strategies, and it can be defined as follows: start by cooperating, then repeat the move the other player made in the previous turn. In simple terms, if the other player cooperated last turn, you cooperate the next, and vice versa. This has provided a partial explanation for the initial evolutionary dilemma (this natural reciprocity is very economical and "encourages cooperation," often under the penalty of mutual destruction) and obviously, while showcasing the substantial "mathematical power" of such an approach, it should not be taken as a "universal key" capable of resolving every situation of this kind. However, I consider it an extremely interesting case, capable of offering valuable insights in a variety of fields, such as social psychology, economics, politics, and even diplomacy.

But let's move away from the intriguing idea of repeated games and return to analyzing the simpler cases of single-situation games. For this purpose, let's consider a game that turns out to be completely devoid of pure dominant strategies; in particular, let's examine a situation that often arises during poker games. For anyone who knows even a little about the game, it shouldn't be surprising that there isn't a single optimal decision that can be taken "always" and independently of every game situation.

Let's therefore consider a situation that might be common during poker games: there are 2,000 dollars in the pot, out of which we have already bet 300. Then, our opponent bets another 1,000, and we have a hand that is not exceptional, but average, and we are the only ones left in the game. We must therefore decide, based on whether our opponent is bluffing or not, whether to call the hand or not. We can outline this situation through the table below:

	Other one is bluffing	Other one isn't bluffing
We call	2000	-1000
We fold	0	300

If we only consider the maximum values in each row, we can immediately notice that there is no value that significantly surpasses the others. In such a situation, therefore, we can resort to a "mixed" strategy, which is something that, at least in theory, has a greater likelihood of success.

That is, we can calculate the "difference value" (DV), which is the difference between the maximum value and the minimum value in each row. We then add this DV to the OTHER row, and the result should give us an estimate of the frequency with which we should choose each option to maximize our expected gains.

But since an example is worth more than a thousand explanations, let's take a detailed look at an application of what has just been said:

- **Line "we call":** maximum value 2000, minimum value -1000. Since subtracting a negative number is equivalent to adding the absolute value of that same number, we have 2000 - (-1000) = 2000 + 1000 = 3000. This value goes in the other line, which is the "we retreat" line.

- **Line "we fold":** maximum value 300, minimum value 0. 300 - 0 = 300. This value goes in the line "we look."

The result is the table:

	Other one is bluffing	Other one isn't bluffing	DV
We call	2000	-1000	300 (≈9%)
We fold	0	300	3000 (≈91%)

What do those values mean, then, in terms of the frequency of our mixed strategy? Simply that, given these bets, it would be probabilistically advantageous for us to fold 3000 times and to call 300 times. In other words, by simplifying the 300/3000 fraction, it would be advantageous to *call once and fold ten times* (or, respectively, call with a 9% probability and fold with a 91% probability). Therefore, in the situation where this choice appears only once, we could act "randomly" based on the given frequency, checking our watch, and deciding to bet only if the second hand (or the face) indicates that we are within the first 6 seconds of the current minute. By doing so, it's guaranteed that you'll bet precisely with the probability of one in ten.

In short, we can say that the formula to follow here is: fold (the pot plus what you need to bet to call) times, and call (what you have already invested in the pot) times. This strategy, being based on a purely probabilistic evaluation, will not guarantee that we win the hand every time, but it will ensure that, under similar conditions, we do so in the majority of cases.

Clearly, everything discussed so far should be taken as *a simplification of a real game situation;* it should not be applied literally! It assumes that you are unable for example, amongst other things, to decipher your opponent's behavior. And that you have not incorporated any other game factors into your evaluations, such as the cards already played, the "quality of your hand," or prior knowledge of your opponent (as in the case of repeated games). However, if the fact that your opponent is known to bluff very often, that you have a straight flush in hand (with a probability of about 1 in 50,000, making it very unlikely to lose the hand), or a statistical analysis of the cards played up to that moment suggests that calling the hand is more likely, you can clearly improve the strategy discussed so far. For example, if you could know or estimate that there is only a 10% chance that the opponent is bluffing, you could take the same table or formula discussed, replace the values in that column with their 10% (a multiplication by 0.1), those in the other column with their 90% (a multiplication by 0.9), and reassess your considerations based on the new values obtained. Consider what has just been said as a simple introduction to both probability theory and the very basic yet valuable idea that *knowledge is power* in situations lacking pure strategies. For all necessary in-depth explorations of probability theory, we will revisit this in a few chapters.

Certainly, someone here might rightly point out that in the case of bluffing, or prison years, the process is greatly simplified by dealing with already well-defined numerical quantities. Therefore, in other cases, creating a table of advantages might be difficult. But to address this objection, let's simply recall what was mentioned at the beginning of the book: in cases where quantities cannot be clearly defined or calculated, we can always assign a *score* in the table corresponding to the benefit we gain in the presence of a specific occurrence.

But before concluding the chapter, let's take another look at a broader topic. In the early 1960s, a mathematician named John Nash, whom you probably know from the film "A Beautiful Mind," formalized what is likely one of the most important mathematical concepts of the last century. Nash realised he was living in a society that was primarily based on the economic-mathematical principles expressed by Adam Smith. According to these principles, maximum prosperity within a group can only be achieved when each individual strives to the fullest to pursue their personal gain.

Nash, however, after years of intense work, proposed a revolutionary theory known as Nash Equilibrium, demonstrating that in many contexts, individual optimization can align with the common good, and that in other situations, individuals acting selfishly can drive the system to a suboptimal state (known as the "tragedy of the commons"). This does not require some form of "sacrificing for the community," nor does it mean completely forgoing individual benefit. Rather, taking into account the distinctions from case to case, this formulation invites us to consider the possibility of a more sophisticated equilibrium, a compromise that can benefit both the individual and the broader context (despite the fact that, and we must highlight it, in most of cases this equilibrium could be exceptionally hard to compute).

A "golden rule" that can be derived from all this (possibly in conjunction with what said regarding the evolutionary advantage of the "tit for tat" strategy) is: every time you have to make a decision, try to look for the "positive equilibrium." Carefully craft (or look for) contexts in which, for most of the relevant players, it would be suboptimal to retreat from a cooperative stance. Even better, think and act in win-win terms. Naturally, this won't always be possible, and indeed, most of the time it will be incredibly complex, but it's possible that if you at least try to broaden your perspective in this direction (once again, without giving up individual benefit, especially where it is necessary for your own "survival"), you might gain an enormous additional advantage in terms of happiness, success, and satisfaction, compared to what an antagonistic or egoistic approach would initially suggest.

Cooperation and a "win-win oriented" perspective seem, after all, to be the only real keys to the future of any human group, from the smallest family to the much larger group inhabiting the entire planet. The complex problems we will face as a species, such as pandemics, climate change, or the need to find new habitable planets, will require a collective effort that transcends the borders of continents and nations, prompting us to think in terms of the entire humanity. This path is the only reasonable way for human societies, and the individuals within them, to achieve the most and the best. And all of this is much more than just a set of trivial rules or rhetoric: it is something real, undeniable, numerical, like the very laws that define the orbits of planets around the sun or the acceleration of objects in free fall: *it is mathematics!*

The Historical Pill

The Roman numeral system (where I represents 1, V represents 5, X represents 10, and so on), although well-known enough to still be used today, was incredibly inconvenient for performing calculations. To overcome this obstacle, the ancient Romans often resorted to using tools like the abacus or simple stones known as "calculi" (from which the modern word "calculation" is derived). The abacus, in particular, was a widely used tool in ancient Rome; it typically consisted of a board with grooves in which pebbles or markers were placed, and each groove or line could represent a power of ten: the first for units, the second for tens, and so on. Through this "small, simple instrument," the Romans were able to carry out all the "necessary" mathematical operations to expand and manage such a vast and complex empire.

The Aphorism

"Mathematics is the place where you can do things that are not possible in the real world."
(Marcus du Sautoy)

Mathematical Oddities

By modeling countless practical situations through game theory, it becomes clear that occasionally making a random choice can sometimes prove more successful than a rational or planned one. This happens because, in many situations, not only does acting predictably expose us to exploitative strategies from opponents, but we also tend to underestimate our own level of predictability. This applies even if the "opponents" are not sentient beings, but any circumstance or context where we intend to achieve a result: any action not already anticipated by the "natural defenses" or the rules of the system in question could be much more effective than the alternatives.

IX - Break, Simplify, and Save Money

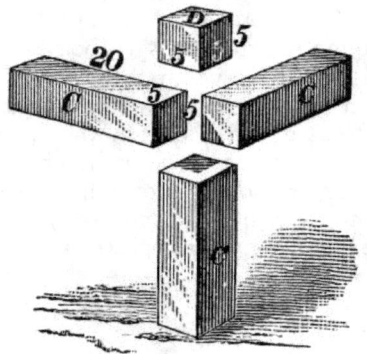

From this chapter begins a comprehensive and detailed discussion dedicated to strategies for quick multiplication and division.

As you will see as you progress through the pages of this book, mathematicians have always paid particular "attention" to these two operations, developing a vast array of rapid calculation techniques dedicated to them. And, as you already know, the very fact of having many possible tools to tackle a problem can give you the freedom to approach it in a playful, creative way while also using the strategic method that suits you best.

Moreover, multiplication is undoubtedly a critical and fundamental operation within "everyday mathematics": just consider its usefulness in any economic transaction, any calculation that requires percentage evaluation, or (as we will see shortly) any strategic decision that requires a basic understanding of probability theory.

But a piece of advice I feel like repeating to you before diving into the heart of these methods is: don't try to "digest" them all at once. Instead, read them little by little, replicate the examples with pen and paper, and gradually determine which ones are more suitable for you.

Now, revisiting a principle already stated in previous chapters, which posits that "the human mind is able to perform many simple operations more quickly and accurately than a single complex operation," we will begin by introducing three extremely useful techniques to transform complex multiplications into sequences of simpler multiplications.

Addends decomposition

The first technique we will examine is the so-called "addends decomposition," which is applicable only to multiplication and useful when dealing with operands composed of two or more digits.

Very simply, this technique consists of two steps:

- One of the operands is broken down into a sum or a subtraction. Typically, this is done by adding its units, plus its tens, plus its hundreds, etc. (for example: 456 = 400 + 50 + 6)
- The multiplication on these parts is performed separately by taking advantage of the distributive property of addition. And so if for example we want to calculate 334 x 456, we can transform this multiplication into

$$(334 \times 400) + (334 \times 50) + (334 \times 6)$$

That is, instead of performing a multiplication between three-digit numbers, which would require a certain amount of time if written in column form, we end up performing a sum of simpler single-digit multiplications, to which we will then need to add an amount of trailing zeros that's depending on the original number (you've refreshed on how it works with multiplications by 10, 100, 1000, right?).

We could have also broken down 334 into 300 + 30 + 4 and thus transform the original multiplication into

$$(456 \times 300) + (456 \times 30) + (456 \times 4)$$

The commutative property of multiplication dictates that if you break down one multiplicand rather than another, the final result doesn't change. Therefore, it will be up to us each time to choose the simplification that leads to more immediate micro-operations.

Warning: breaking down into units, tens, and hundreds is just an option, and a multiplicand can be decomposed into any sum or subtraction that helps simplify the final operations. For example, if we want to multiply 44 x 7, we can alternatively decompose 7 into (10 - 3), so

$$44 \times 7 = (44 \times 10) - (44 \times 3)$$

And this last one is a criterion that greatly simplifies our lives whenever we have to multiply numbers close to a multiple of 10. Indeed, for example, if we find ourselves multiplying 400 x 59, instead of breaking down 59 into (50 + 9), it might be more convenient to calculate it as (60 - 1), and therefore:

$$400 \times 59 = 23{,}600$$

It's clear that, as someone here might easily notice, even after applying these "precautions," these operations might not always be easy to perform mentally. However, the advice remains the same: first of all, consider that these techniques are designed to simplify arithmetic calculations in general, whether they are mental, written, or performed with the aid of external tools. If you want to set yourself the goal of using them to enhance your cognitive-level mathematical speed, then try starting with simple operations and progressively train yourself with increasingly complex ones. Learn to keep in mind only what's strictly necessary, but most importantly... discard any technique that you feel is not particularly suited for you.

Factor decomposition

Let us now move on to the second type of decomposition: *factor decomposition*, which can also be used for division and requires some additional context.

First of all, let's remember that factoring a number means breaking it down into a multiplication of smaller whole numbers. I can also apply to a number the so-called *factoring into prime factors*, which obviously still means breaking it down, but into a multiplication of *prime* whole numbers. And, as you surely know, a prime number is an integer that is divisible only by one and itself.

For example, I factor 16 when I say it is equal to 4 x 4. I break it down into prime factors when I say it is equal to 2 x 2 x 2 x 2.

Now consider the advantage of being able to factor a number, even better if into prime factors. For example, 88 x 16 = 88 x (2 x 2 x 2 x 2). The associative property of multiplication tells us that we can perform these operations in any order, so multiplying by 16 will be equivalent to the much faster operation of doubling our original number four times.

A similar thing happens if we want to divide a number by 16, for example 256. We would have that 256 / 16 = 256 / (2 x 2 x 2 x 2). That is, dividing a number by 16, for example, is equivalent to halving a number four times. Similarly, if we wanted to divide a number by 96, and 96 = 2 x 2 x 2 x 2 x 2 x 3, we would first need to halve the number five times and then divide it by 3.

Certainly, factorization is definitely not an immediate operation, especially in the case of large numbers, and that's why it's necessary to learn, or at least remember, the so-called *divisibility rules*. Let's say, for example, that we want to factorize 256 and proceed step by step:

- Knowing that 256 is divisible by 4 allows us to immediately determine that 256 can be broken down into "4 multiplied by something."
- We get that "something" by doing 256 / 4 = 64. Therefore, 256 = 4 x 64. I hope, moreover, that this step and the previous one are clear to everyone: they "obviously" stem from the fact that multiplication is the arithmetic inverse operation of division.
- We (should) know that 4 is divisible by 2, and that it equals 2 x 2. Similarly, 64 is divisible by 2, and being twice 32, it equals 32 x 2. Therefore, by substituting 4 and 64 with their factors, we have that 256 = 4 x 64 = 2 x 2 x 2 x 32.
- And so on, stopping the decomposition when I arrive at having all prime factors or when the factorization allows us to operate "quite simply."

Clearly, from this process, it becomes evident that it is not advantageous to decompose a number further if it makes the operation too complicated. And precisely for this reason, knowing the most useful and practical divisibility criteria allows us not only to operate with greater simplicity but also to understand when it is appropriate to discard the method.

Divisibility by 2 rule: This is probably the simplest and most well-known: a number is divisible by 2 if it is even, meaning if and only if its units digit is 0, 2, 4, 6, or 8.

Divisibility by 3 rule: A number is divisible by 3 if the sum of its digits results in a multiple of 3. For example, let's consider 476. 4 + 7 - 6 equals 17. If 17 is a multiple of 3, then 476 would also be, and to determine this, you can apply the same rule again: 1 + 7 equals 8. Since 8 is not a multiple of 3, neither is 476.

Divisibility by 4 rule: A number is divisible by 4 if its last two digits are 00 or a number divisible by 4. For example, take a number like 56,000,932. I can immediately say that it is divisible by 4 because it ends with "32" and 32 is divisible by 4.

Divisibility by 5 rule: A number is divisible by 5 if its units digit is 0 or 5.

Divisibility by 6 rule: A number is divisible by 6 if it meets the divisibility rule for 3 and it is even.

Divisibility by 7 rule: A number is divisible by 7 if you double the last digit of the number, subtract it from the rest of the number, and the result is zero or a multiple of 7. This can be a bit tricky, so let's reserve an example by trying to determine if the number 397 is divisible by 7. In this case, to verify divisibility, we need to double the last digit of 397, which is 7, giving us 14. Next, we subtract this doubled value from the rest of the number, 39, resulting in

$$39 - 14 = 25$$

Finally, we check if 25 is divisible by 7. Since 25 is not a multiple of 7, we conclude that 397 is not divisible by 7.

Divisibility by 8 rule: A number is divisible by 8 if its last three digits are divisible by 8. Alternatively, you can take the third last digit (from the right), double it, add it to the second last digit, double that result, and add it to the last digit. If the final result is a multiple of 8, then the original number is also divisible by 8.

So, for example, if you have 865,341, you can apply the rule as follows: take 3 (the third last digit), double it to get 6, add it to 4 to get 10. Then, double 10 to get 20, and add 1 to get 21. Since 21 is not divisible by 8, neither is 865,341.

Divisibility by 9 rule: A number is divisible by 9 if the sum of its digits is divisible by 9.

Divisibility by 10, 100, 1000 rule: A number is divisible by 10, 100, 1000, or any other power of 10 if it ends with a number of zeros equal to those in the power considered (simple as that).

Divisibility by 11 rule: Starting from the right moving to the left, add the digits in the even positions of the number you want to examine. Then add those in the odd positions. Subtract the smaller sum from the larger sum, and if the result is 0, 11, or a multiple of 11, then the number in question is also divisible by 11.

For example, if we take 8,291,778, we can determine it is divisible by 11 because:

$$(8 + 7 + 9 + 8) - (7 + 1 + 2) = 32 - 10 = 22$$

You will also find later on that "discovering" a number is divisible by 11 allows you to access a very quick and effective calculation tool.

Divisibility by 12 rule: A number is divisible by 12 if it meets both the divisibility rules for 3 and for 4.

Divisibility by 14 rule: A number is divisible by 14 if it is divisible by 7 and is even. From this, it should also be easy to understand when a number is divisible, for example, by 18 or 22.

Divisibility by 25 rule: A number is divisible by 25 if it has at least two digits, and the last two digits on the right are 00, 25, 50, or 75.

Expressions decomposition

This technique involves breaking down a number into a set of multiplications and divisions (an "expression") that theoretically could simplify the original operation.

For example, 50 is equal to 100 / 2. Therefore, multiplying a number "a" by 50 can be approached by "following the expression" that is, by first multiplying "a" by 100 and then halving the result. Conversely, division by 50 can be carried out by "reversing the expression," meaning exchanging each multiplication with a division and vice versa, thus dividing "a" by 100 and then doubling the result.

There is no precise criterion for breaking down a number into an equivalent expression. Let's say it can be done when you've trained your eye a bit to recognize whether a number is a factor of 10 or one of its multiples, and it's not worth applying other calculation strategies. For example:

- **5 = 10 / 2.** Therefore, multiplying by 5 is equivalent to multiplying by 10 and then halving. Dividing by 5, on the other hand, involves reversing the expression, which means dividing by 10 and then doubling.
- **25 = 100 / 4.** Therefore, multiplying by 25 is equivalent to multiplying by 100 and halving twice. Dividing by 25 requires us to reverse the expression, thus it equates to dividing by 100 and doubling twice.
- **75 = 300 / 4.** Therefore, multiplying by 75 is equivalent to multiplying by 100, then tripling, and finally halving twice. On the other hand, dividing is equivalent to dividing by 100, then doubling twice, and finally dividing by 3.
- **250 = 1000 / 4.** Therefore, multiplying by 250 is the same as multiplying by 1000 and halving twice. Division is equivalent to inversion.

You could even combine the *expressions decomposition* with the *addend's decomposition* and solve, for example, the multiplication by 26 very simply. In fact, 26 is equal to (100 / 4) + 1. Thus, you find that it is possible to multiply a number "a" by 26 by multiplying "a" by 100, halving twice, and adding the same "a."

I will say once again that this is a process that greatly simplifies the initial operations but, at the same time, may require a lot of "eye" and intuition. Therefore, whenever you find that trying to use it is complicating things too much for you, just let it go and rely only on the first two.

Tackle even the slightly more complicated things, you'll now see a very useful quick multiplication and division table, compiled by implementing all types of decomposition we explained so far. Rather than memorizing it, my suggestion is to try to understand it, so that you can extend it to larger numbers and thus easily use its criteria to solve a great number of operations.

—-

Multiplication by 4: Double twice.

Division by 4: Halve it twice.

Multiplication by 5: Multiply by 10 and then halve, or vice versa.

Division by 5: Divide by 10 and then double, or vice versa.

Multiplying by 6: Double and then triple, or vice versa.

Division by 6: Halve it and then divide by three, or vice versa.

Multiplication by 7: Multiply by 10 and subtract three times the original number (here note that it is not possible to create any rule for division by 7, because *addends decomposition* is not applicable to division, *factor decomposition* is not feasible since 7 is a prime number, and a *expressions decomposition* is not achievable.)

Multiplication by 8: Double the original number three times.

Divide by 8: Halve the original number three times.

Multiplication by 9: Multiply by 10 and subtract the original number.

Division by 9: Divide twice by 3 (note here that we've used one type of decomposition for multiplication and another type for division).

Multiplication by 11: Multiply by 10 and add the original number.

Multiplication by 12: Multiply by 10 and add the original number twice.

Division by 12: Halve twice, then divide by 3.

Multiplication by 13: Multiply by 10 and add the triple of the original number.

———

Multiplication by 14: Multiply by 7 and then double it.

Division by 14: Halve it and then divide by 7.

———

Multiplication by 15: Multiply the original number by 10 and add half of it.

Division by 15: Divide by 3 and then by 5.

———

Multiplication by 16: Double four times.

Division by 16: Halve it four times.

———

Multiplication by 17: Multiply by 20 and subtract the triple of the original number.

———

Multiplication by 18: Multiply by 20 and subtract twice the original number from the result.

Divide by 18: Halve, then divide twice by three.

———

Multiplication by 19: Double the number, then multiply by 10, and subtract the original number from the result.

———

Multiply by 20: Double and multiply by 10.

Divide by 20: Halve it and divide by 10.

———

At this point, you could use these strategies to perform a "mathematical trick" that will be really useful in managing your personal finances: understanding the approximate annual cost of an expense you incur every day. How do you do it? Well, it's actually quite simple:

- Consider your daily spending, let's say 5 dollars for eating out during lunch at work instead of bringing homemade food.
- Multiply it by the days of the week during which you actually spend this money. In this case, let's say it's 5 days = 25 dollars. This is your weekly cost.
- Now all you need to do is multiply the weekly cost by 52 to obtain your annual expense. How to do it? Well, it's simple, we can split 52 into 50 + 2 and thus see that 25 x 52 = 25 x (50 + 2) = 25 x 50 + 25 x 2.
- Multiplying by 50 at this point should be super simple: since 50 = 100 / 2, you just need to multiply by 100 (thus adding two zeros) and then divide by half = 25 → 2500 → 1250.
- Now the approximation for the expense is already good enough to understand how those innocent-seeming 5 dollars on weekdays can add up by the end of the year. If we really want to be precise, we can add the remaining two weeks and calculate 25 x 2 = 50, which when added to that 1250 equals 1300. Ouch! Better to prepare food at home, isn't it?

Similarly, think about how much you spend on your monthly subscriptions like the gym, streaming platforms, or food delivery services. If your gym membership, for example, costs 30 dollars a month, you already know how to multiply by 12: multiply by 10 and add double the original number. So, 30 x 10 = 300 and 30 x 2 = 60. Adding the two results gives us 360 dollars a year, and that's just for the gym!

Because from here we could also continue with other expenses like gasoline, parking, newspaper subscriptions, streaming services, and so on, and by adding up all these amounts, we could immediately understand where a good part of our salary goes by the end of the year. It's important to clarify that everything discussed so far should certainly not encourage an extreme ultra-minimal lifestyle where we give up all conveniences, subscriptions, or sports activities (after all, you probably didn't need this book to realize, for example, how important sports and physical exercise are for your physical and mental health). Rather, it is about truly grasping the mechanism explained a few pages ago, where our brain "struggles" to think quantitatively. Driven by stimuli perceived as more attractive, "mistakenly relevant," and fascinating, this same brain leads us to say "yes" to a plethora of small daily, weekly, and monthly expenses, which, under the illusion that their benefits outweigh the overall cost, end up seriously damaging our finances, away from our ability to notice. In this regard, as well as with the practical suggestion to clean up unnecessary recurring expenses every 6-12 months, I would like to conclude this chapter by emphasizing how "criminal" it almost is the way our educational system remains silent on what is perhaps the most important value mathematics can add to our daily lives: the way logical reasoning and the quantitative vision associated with it are the best means to provide us with an objective, rational, "solid" view of reality. This not only greatly enhances our ability to influence the world, but it also gives us true decision-making freedom, far from the manipulative attempts of those who try to alter this reality to turn us into puppets, soldiers of this or that cause, or merely "soulless consumers."

The Exercise

Try calculating the annual impact of your expenses for anything that represents: 1) A fixed recurring cost, like the gym or a streaming service subscription, and 2) Expenses you would classify as "auxiliary," meaning they are neither mandatory nor essential for your health, business, or family. Do this separately for each type of periodic expense, and then calculate the total. After that, simply assess whether it might be worth "cutting out" or perhaps replacing something with a more cost-effective equivalent, especially if its cost/benefit ratio isn't particularly advantageous.

The Historical Pill

A legend, probably dating back to the sixth century AD, tells that the inventor of the game of chess, Sissa ibn Dahir, when asked to name his reward for the invention, quantified it as follows: "Place one grain of wheat on the first square, and double the amount for each subsequent square." If this request had been accepted, it would have required more than 18 quintillion grains of wheat, a quantity far greater than what could be obtained after cultivating the entire planet with wheat and setting aside several months of harvests. This story is often used to demonstrate the completely counterintuitive nature of exponential growth.

The Aphorism

"The mathematical sciences display order, symmetry, and limitations in a particular way, and these are the highest forms of beauty."
(Aristotle)

Mathematical Oddities

For those who are not familiar with it, the Pareto principle, also known as the 80/20 rule, posits that for many events, approximately 80% of the effects come from 20% of the causes. This empirical rule has been observed in a wide variety of contexts: in economics, for instance, 80% of wealth is held by 20% of the population; in software, 80% of bugs can usually be found in 20% of the source code; in health, 80% of your ailments may stem from 20% of potential risk factors, and so on.

The Pareto principle is indeed a concept that, despite being much more than just "acknowledged" today, remains, in my opinion, underestimated in the analysis of problems and solutions. It is extremely powerful for ensuring a correct focus of efforts and resource allocation.

X - The World of "Golden Numbers"

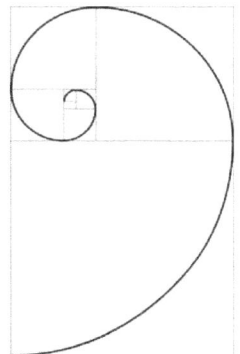

Let's now delve into the realm of numerical "curiosities" and begin exploring a unique class of numbers I refer to as 'golden numbers.' These are numbers with particular intrinsic harmony, granting them unique properties or simplifying certain calculations. Although many of these properties may have minimal practical applications, some could be useful for performances in rapid calculation or mathematical tricks. Moreover, when practical applications are absent, my goal is to try intriguing you with the beauty these numbers exhibit through symmetry and harmony. This beauty, after all, is everywhere around us, as it can often be seen in the harmonies present in nature, the patterns in historical social and economic phenomena, or the intricate logic in algorithms and architectures. But let's begin!

$$\varphi$$

Despite having arbitrarily decided to label all the numbers I'll discuss in this chapter as "golden," the truth is that the only officially recognized "golden number" is the one expressible as (1

+ √5)/2 = approximately 1.6180339887, often denoted as "phi" (φ). This number is arguably one of the most "famous" mathematical constants of all time.

This number is linked to the Fibonacci sequence, a series of numbers where each number is the sum of the two preceding ones (1, 1, 2, 3, 5, 8, 13, 21, 34, 55, etc.). As you progress through the sequence, the ratio between a number and its predecessor increasingly approaches the golden ratio. It was also "revered" in ancient Greece for the way it frequently appeared in nature (for example, the arrangement of leaves on many plants often follows the Fibonacci sequence, allowing each leaf to receive maximum exposure to sunlight and rain. Similarly, the arrangement of seeds in a sunflower follows the Fibonacci sequence, enabling the plant to pack the maximum number of seeds in a limited space). Additionally, using the visual proportion of 1:1.6180339887 in painting, sculpture, and architecture provided a great sense of harmony and balance. However, its extensive range of practical applications didn't cease in ancient Greece. The dimensions of many credit cards and business cards closely resemble the golden ratio, as do the designs of many technological products like televisions (think of 16:9), computer monitors, and smartphones. The golden section also repeatedly appears in widely varying fields of human endeavor, such as music (for instance, the idea of placing the chorus or the climax of a composition at a point with a golden ratio relative to the total duration of the piece might be particularly "pleasing"), or finance (in financial analysis, for example, tools like "Fibonacci retracements" are sometimes used to predict subsequent corrections when analyzing significant price variations. Though I wouldn't particularly recommend it, so consider it more as a "case study").

A practical curiosity is that the ratio between two consecutive numbers in the Fibonacci sequence closely approximates the conversion ratio between kilometers and miles. So, if we wanted to, we could add to our "mathematical arsenal" the somewhat approximate technique: 1) When you need to convert from kilometers to miles, go one step back in the Fibonacci sequence (for example, 55 kilometers equals 34 miles), 2) When you need to convert from miles to kilometers, go in the opposite direction, and 3) If, for any reason, the number you're interested in is not in the Fibonacci sequence, you could still apply a very simple principle of *interpolation or approximation*. For example, if you need to convert 60 kilometers to miles, which is not a number in the Fibonacci sequence, you could consider it as the sum of 55 and 5 (two numbers belonging to the sequence) and then add the corresponding miles, 34 and 3, to get an estimate of 37 miles. This approximation would indeed be extremely close to the exact value, which is 37.28 miles.

Finally, an everyday application worth discussing is the "rule of thirds" in photography, though it is an approximation. To apply this rule, the photograph is imagined as being divided by a 3x3 grid. Then, placing the main subject along the lines where this grid intersects is believed to enhance the composition's aesthetic appeal.

Of course, like all "universal rules" that attempt to decode the aesthetic canons of any given art form, this is a principle that professionals, quite rightly, will declare as "made to be broken." Nevertheless, I find it extremely fascinating how this "constant of nature" keeps reappearing, as if it were intrinsically woven into the very "DNA" of the universe, constantly encouraging us to create and experiment based on its very harmonies. How would you view, for example, the idea of creating your own personalized "Pomodoro Technique" (a productive technique that involves alternating periods of work and rest following intervals defined by a timer) that follows, perhaps in descending order, a number of minutes corresponding to the Fibonacci sequence?

3

Throughout human history, the number 3 has often been attributed symbolic and esoteric meanings. It has frequently been considered the perfect number par excellence, even though it has nothing to do with the mathematical definition of a "perfect number." Its recurrence, not coincidentally, seems to permeate countless fields of existence: there are (depending on the models) three spatial dimensions in which we move, three primary colors, three types of fundamental atomic particles, and so on.

Beyond, however, the search for meaning and symbolism behind a certain number which, while undoubtedly fascinating, often risks crossing into "magical thinking" and forcing reality to match our desire to find patterns (while remaining "pretty convinced" that the recurrence of the golden ratio might be an exception to this principle), let's attempt to rediscover a "more verifiable harmony" of the number 3 by examining a method to instantly calculate the squares of numbers composed solely of 3s. In particular, to perform this operation instantly you can simply:

- Write as many 1s as there are 3s in the number to be squared, minus one.
- Write a 0.
- Write as many 8s as the number of 1s you wrote in the first point.
- Write a 9.

Here, for instance:

$$33^2 = 1089$$
$$333^2 = 110{,}889$$
$$3333^2 = 11{,}108{,}889$$
$$33333^2 = 1{,}111{,}088{,}889$$

And so on.

The section about "3" concludes here. Sure, maybe this is a trick with highly limited practical use, but it remains a small way to demonstrate, once again, the sophisticated beauty of the realm of numbers. Moreover, if we really want to take an additional "leap of imagination," what we've seen so far could, for example, be particularly appealing to teach a child just starting with mathematics, to consolidate their mental calculation skills, help them develop their ability to recognize numerical patterns, and, why not, allow them to "impress" when they can show how easily they calculate squares of numbers with so many digits.

9

The "closeness" of 9 and numbers composed entirely of 9s to the powers of 10 has always imparted a fascinating harmony to the results of operations performed on them. It is precisely for this reason that we can exploit this "unusual harmony" to perform various types of operations extremely quickly. For example:

"Golden case" A: Rapid Addition

To add to any number "a" another number composed of the same number of digits, but made up only of 9s, just subtract 1 from "a" and place this "1" to the left of the number:

$$88 + 99 = 187$$
$$543 + 999 = 1542$$
$$2,342 + 9,999 = 12,341$$

This "very simple" trick, unlike the previous ones, actually offers a lot of interesting practical applications, as it allows you to quickly add a number to another that is very close to 99, 999, 9999, etc.

If, for example, you wanted to calculate 567 + 997, you could apply this technique and then, considering that 997 is 2 less than 999, subtract 2 from the result. It might seem trivial, but it's extremely quick and effective.

"Golden case" B: The Return of the Sutra

The properties of the number 9 make it extremely easy to multiply numbers composed solely of 9s by any other number, as long as they have the same number of digits. In particular, to perform this operation you can:

- Subtract 1 from the number you want to multiply by the series of 9, and place this difference to the left in the partial result.
- Apply the Sutra "All from 9, the last from 10" to the same number as we discussed in the chapter on Vedic Mathematics, and place it after the previous result (remember that it's the Sutra which involves subtracting each digit of the number from 9, except the last one which is subtracted from 10, right?).

Let's now go through a few examples to demonstrate how simple this method is:

89 x 99 = 88 (89 - 1) is the left part of the result, 11 (Sutra on 89) is the right part = 8811

768 x 999 = 767 moves to the left, 232 which is the result of the Sutra on the right = 767,232

3451 x 9999 = 3450 goes to the left, 6549 which is the result of the Sutra to the right = 34,506,549

"Golden case" C: Ultra-Multiplication Table

By multiplying 9 by any number consisting of identical digits, the result can be obtained, ultra-rapidly, in the following way:

- Multiply 9 by the repeated digit. Place the tens on the left and the units on the right.
- Now, in the middle, insert as many nines as there were digits in the original number minus one.

So, for example:

9 x 666 = (6 x 9 equals 54, we will put the two digits on the sides and in the middle we will put two 9s) **= 5994**

9 x 7777 = 69993

9 x 88888 = 799992

"Golden case" D: Instant Division

Another interesting property of this "golden" number is that, by drawing from another Sutra of Vedic Mathematics, it is possible to divide any number by 9 almost instantly.

In the case of a two-digit number "a" (so if, for example, you want to calculate 68 / 9), proceed as follows:

- The first digit of "a" is the partial result (for example, 68 → 6)
- Add the first and second digits of "a." This will be the remainder (for example, 6 + 8 = 14).
- If the sum that results in the "remainder" is less than 9, stop here: the partial result will be the final result. However, if the sum you calculated in the previous step is equal to or greater than 9 (just like with 68), then increase the partial result by 1 (the initial 68 → 6 → plus 1 = 7), while the actual remainder is obtained by subtracting 9 from what you previously had (so, since 14 - 9 = 5, in this case, you have the result as 7 with a remainder of 5).

If the number "a" has 3 digits (for example, 327 / 9), then proceed as follows:

- Take the first two digits of "a" and add its hundreds digit (so 32 + 3 = 35). This will be the partial result.
- Add the three digits of "a" together (thus 3 + 2 + 7 = 12). They will be the remainder.
- If the sum that results in the "remainder" is less than 9, stop here. However, if that sum is greater than or equal to 9 (as in this case), you must subtract 9 from the remainder until it's less than 9 (and so, in this case, 12 - 9 = 3) and add "1" to the result for each 9 you have subtracted (therefore, in this example, since you subtracted one single 9, the result from 35 will become 36).

"Golden case" E: "magic" squares

Let's conclude this long discussion on the "golden cases for nine" by explaining how to instantly calculate the square of any number composed solely of "9s":

- Write as many 9s as there are in the number to be squared, minus one.
- Write an 8
- Write as many 0s as the number of 9s you wrote in the first item.
- Write a 1

So, here it is:

$$99 \times 99 = 9,801$$
$$999 \times 999 = 998,001$$
$$9999 \times 9999 = 99,980,001$$
$$99999 \times 99999 = 9,999,800,001$$

And so on. Even this last case, just like in the "3" one, may seem relevant only from a playful point of view; yet, in a few chapters, we will see that it can be integrated with an additional rule to allow us to quickly calculate the squares of numbers very close to powers of 10.

11

The number 11, and more generally, all numbers composed solely of "1"s, are numbers whose harmony in mathematical results, much like it happened with number nine, derives from their proximity to powers of 10. However, the curiosities surrounding the number 11 are not limited to arithmetic properties. In geometry, for instance, the hendecagon, a polygon with 11 sides, has always presented countless "mathematical challenges" to those who have ventured into this field. In fact, if we try to construct a regular hendecagon, meaning one with all equal angles and sides, using only a straightedge and compass, as has been done for centuries in the Euclidean tradition, we face an insurmountable obstacle. The construction of a regular polygon is only possible if the number of its sides follows specific rules (in other words, for those inclined to delve deeper, if and only if the greatest odd prime factor of that number is 1, or a product of distinct Fermat primes); as such, the number 11 does not fall within these rules, making the construction of such a polygon impossible without resorting to approximations or more advanced mathematical techniques, like spherical trigonometry and non-Euclidean geometries.

However, returning to more distinctly arithmetic and straightforward techniques, let's examine for a moment the properties that multiplication by 11 offers in terms of quick calculation. A first, simple technique to apply here would be one that uses the *addends decomposition* and therefore consists of multiplying a number by 10 and adding the number itself. But in reality, there is another even more refined technique and, in some cases, even quicker:

- Take the first and last digit of the number you want to multiply by 11 and place them as the first and last digit of the result.

- Add the digits of the number you want to multiply by 11, two by two, from left to right. Then, insert between the two outermost digits, again from left to right, the results of these sums. If any of these sums result in a number greater than 10, write only the units digit and carry the tens digit over to the leftmost digit.

Since this method is much easier to perform than to explain, here are some examples:

- **11 x 23** = Place the 2 on the left and the 3 on the right, resulting in a partial answer of 2_3. By adding together the digits of 23 you get 5. You can place it in the middle and there you have the result of the multiplication: 253. Super easy, right?

- **11 x 387** = Place the 3 on the left and the 7 on the right. So you will have a partial result of 3_7. By adding the digits two by two, you have that 3 + 8 = 11. Write 1 and carry the 1 over to the left, making the partial result 41_7. Then, 8 + 7 = 15 → 41_15_7. Keep the 5, carry 1, and get 4257.

- **11 x 4709** = Place the 4 on the left and the 9 on the right, resulting in a partial outcome of 4_9. Adding the digits two by two, You will have 4 + 7 = 11. Partial result 51_9. 7 + 0 = 7, and partial result 517_9. Finally, 0 + 9 = 9, and hence the final result is 51,799.

Let's then conclude the subject with a very simple technique for calculating the square of numbers made up entirely of "1"s:

- *Count.* Yes, exactly, starting from 1, count by writing digit by digit the positive integers (for the less attentive: 1, 2, 3, etc.) and continue until you have on the sheet a number of digits equal to those of the number composed solely of 1s whose square you are calculating.

- Once you have reached that precise number of digits, continue counting and writing, but this time in reverse until you reach 1 again. Once you reach 1, the procedure is complete and you have your result.

Some examples? Here they are:

$$11^2 = 121$$
$$111^2 = 12{,}321$$
$$1111^2 = 1{,}234{,}321$$
$$11111^2 = 123{,}454{,}321$$

A homework assignment? Sure, probably no one out there has ever truly loved homework, but if this topic interests you, try answering this question: how does this rule change if the "1s" are more than nine?

37

37 multiplied by 3 gives us 111. This implies that such a number can be instantaneously multiplied by a multiple "a" of 3 (up to 27) and the result will always be a number composed entirely of "a / 3." So, for example:

$$37 \times 12 = 444 \ (12 \text{ divided by } 3 = 4)$$
$$37 \times 18 = 666 \ (18 \text{ divided by } 3 = 6)$$
$$37 \times 27 = 999$$

This also means that any three-digit number with all identical digits, such as 555 or 777, when divided by the sum of its digits, will always give us 37. The properties of this number don't exactly have any major practical applications, but it should be intuitive: as the value of these "golden numbers" increases, it becomes less likely that you'll encounter them as a component of a practical problem in everyday life. Therefore, with a few rare exceptions, the rest of the chapter should be seen as "pure mathematical entertainment." Continue if interested, and feel free to skip to the next chapter if not.

143

For those who have decided to stay (kudos to you!): let's move on to 143. This number has a very special property: it can be multiplied by another three-digit number "a" simply by forming the number "aa" by "placing" two copies of "a" side by side and then dividing "aa" by 7.

But let's explain it better, as usual, through an example and assume we want to calculate 143 x 887. To perform the multiplication ultra-quickly using this property, we could simply "construct" the number 887,887 by placing two copies of the original multiplier together and then divide it by 7. And if we are "trained enough" to quickly perform such a single-digit division, the operation might be much more immediate for us than the original three-digit multiplication; after all, 143 is a number that lends itself very well to some playful demonstrations of mathematical speed.

666

Actually, it would be more appropriate to discuss numbers made up solely of "6," but the "ominous" symbolism behind this number, which originates, for those who may not know, in the Book of Revelation in the New Testament, was too strong for me not to "provocatively" choose to include it in this form.

However, returning to more "practical" matters, let's analyze for a moment the fact that numbers composed solely of the digit 6 (and greater than 6) have a rather curious property: just like those composed solely of the digit "1" or "9," their square can be calculated very quickly. It is sufficient to:

- Write as many 4s as there were 6s in the original number, minus one.
- Write a 3.
- Write as many 5s as there are 4s.
- Write a 6.

So, for example:

$$66^2 = 4356$$
$$666^2 = 443{,}556.$$
$$6666^2 = 44{,}435{,}556.$$
$$66666^2 = 4{,}444{,}355{,}556.$$

...

One last curiosity about the "6" that might be interesting to explore (even though it is probably known to many of you) is the fact that 6 belongs to the group of so-called "perfect numbers," that is, those numbers equal to the sum of their divisors. For instance, 6 is perfect because it is equal both to 1 + 2 + 3 and to 1 x 2 x 3.

Here too, the symbolism associated with perfect numbers is so strong that Saint Augustine wrote: "6 is a perfect number in itself, and not because God created all things in 6 days. On the contrary, the opposite is true: God created all things in 6 days precisely because this is a perfect number." Moreover, 28 is also a perfect number and, although it does not have particular mathematical properties, it has always held a strong symbolic and religious significance, being, for example, the number of days in the lunar month in the Hebrew calendar.

Let it be clear once again: except for the golden ratio, whose recurrence might have specific reasons (it could be a "sufficiently economical" constant for the self-optimization of some physical systems, for example), I firmly believe that the universe is so complex that it is virtually impossible not to find numerical patterns of all kinds here and there if one is determined to do so. However, beyond the obvious biases that drive us to create certain stories, I still find that these narratives, these myths, are a fundamental part of what we represent as a species and, therefore, I firmly believe that studying them can help us gain a deeper understanding of many of the complex phenomena that animate the human world.

1089

Take any 3-digit number, as long as the hundreds digit is at least 2 greater than the units digit. For example: 542, 890, 341. Not: 101, 615, 789. Now subtract from it the number obtained by swapping the units with the hundreds, like for 542: 542 - 245 = 297. Next, take the result and add the number obtained by swapping the units with the hundreds again. So, 297 + 792. You got 1089, right? This sequence of operations always results in 1089, no matter which 3-digit number you start with, which is why this number can be counted among "magic numbers," in the sense of representing a small but valuable "magician's tool." Have a friend perform these calculations and try to predict the result: it's always a trick that can have quite an effect (... provided it's done only once, of course.)

2025, 3025, 9801

These three numbers have been classified as "golden" numbers because their square root can be calculated almost instantly. The "secret" here is simply to divide them into two groups of two digits and add them. So:

$$\sqrt{2025} = 45 = 20 + 25$$
$$\sqrt{3025} = 55 = 30 + 25$$
$$\sqrt{9801} = 99 = 98 + 1$$

3367

The number 3367 has properties very similar to those of 143. Specifically, to find the result of 3367 multiplied by any two-digit number "a," you can simply create the number "aaa" by placing three "copies" of "a" together and then dividing it by 3. This makes it another "wildcard" number for instant math demonstrations.

So, for example:

$$3367 \times 98 = 329{,}966$$
$$3367 \times 55 = 185{,}185$$

6174

The number 6174 is known as the "Kaprekar constant." If you take any four-digit number (with different digits), arrange the digits in descending order (yes, the largest first), subtract the number formed by the digits in ascending order (the inverse of the previous), and then repeat this process, you will always "sooner or later" reach 6174. Let's take, for example, 8971, whose equivalent sorted in descending order would be 9871, and the inverse 1789:

9871 - 1789 = 8082
(now let's do the same with 8082)
8820 - 0288 = 8532
8532 - 2358 = 6174

The fact that these sequences allow you to reach 6174 in a (pseudo -it could in theory be calculated-)random number of steps might theoretically make it a bad "numerical base" for truly impactful magic tricks. However, in many magic tricks, the theatrical presentation often takes precedence; you could, for instance, lead someone through the subtractions as illustrated so far, stop them suddenly, and then add some "sleight of hand" trick where you pretend to pull out a note with the number 6174 from their ear or pocket.

I'll also reveal one last "magician's secret" to you: the exact same property of 6174 is shared by the number 495, which could likely subject the main "victim" of your magic trick to a slightly less tedious sequence of subtractions.

37.037

37.037 has properties very similar to 37. That is, when multiplied by 3 it gives us 111,111, and because of this, it can be instantly multiplied by any multiple "a" of 3 up to 27, and the result will always be a number composed entirely of "a / 3." Thus:

37,037 x 6 = 222,222
37,037 x 15 = 555,555
37,037 x 24 = 888,888
37,037 x 27 = 999,999

142.857

If you multiply 142,857 by any number between 2 and 6, the result will always retain the same digits as 142,857, but they will reappear "rotated." In fact:

$$142{,}857 \times 2 = 285{,}714$$
$$142{,}857 \times 3 = 428{,}571$$
$$142{,}857 \times 4 = 571{,}428$$
$$142{,}857 \times 5 = 714{,}285$$
$$142{,}857 \times 6 = 857{,}142$$

Furthermore, in case anyone out there is interested, if you multiply the number by 7, the result is 999,999.

12,345,679

This number, which is simply the sequence from 1 to 9 without the 8, has properties very similar to 37 and 37.037. That is, if you multiply it by a number "a" that is a multiple of 9 and less than 90, it results in a number composed of nine identical digits each equal to "a / 9." So, for example:

$$12{,}345{,}679 \times 9 = 111{,}111{,}111$$
$$12{,}345{,}679 \times 36 = 444{,}444{,}444$$
$$12{,}345{,}679 \times 54 = 666{,}666{,}666$$
$$12{,}345{,}679 \times 63 = 777{,}777{,}777$$
$$12{,}345{,}679 \times 81 = 999{,}999{,}999$$

1,016,949,152,542,372,881,355,932,203,389,830,508,474,576,271,186,440,677,966

I doubt you'll ever find yourself dealing with the number 1,016,949,152,542,372,881,355,932,203,389,830,508,474,576,271,186,440,677,966. However, if one day a king promises you his kingdom in exchange for multiplying it by 6 within three seconds, just remove the last digit from the units place and move it to the front. Yes, if 9 and 11 were the most practically useful golden numbers, this is probably the most useless of all.

The Historical Pill

The symbol for the true "0" was imported into the West from the Arab world only starting from the year 1000. Before then, the concept was unknown in Europe, sometimes even considered "taboo" because of its connection with the idea of "nothingness" or "emptiness." In other cultures, it wasn't used simply because, as we saw a few pages ago, numbers were often represented through pebbles, and therefore no pebble corresponded to no number. The invention of zero as we know it today seems to date back to India, between the fourth and sixth centuries AD.

To be clear, this concept already existed before, as can be seen, for example, in some Babylonian clay tablets dating back to 1800 BC. However, India's true innovation was in treating zero as a number in its own right, with its own symbol and mathematical rules, which had an incredibly profound impact on mathematical progress itself.

The Aphorism

"Mathematics is like poetry; it is a completely logical language, a direct path to universal intellect."
(René Descartes)

Mathematical Oddities

In addition to "perfect numbers," there are plenty of "quirky classifications" for certain categories of numbers. Narcissistic numbers, for example, are numbers that are equal to the sum of their digits each raised to the power of the number of digits raised to the power of the number of digits, like 153 = 1 cubed + 5 cubed + 3 cubed. "Twin primes" are prime numbers separated by just one even number, such as 5 and 7 or 11 and 13. There are also "cousin primes," which differ by 4 (e.g., 7 and 11), and "sexy primes," which differ by 6 (e.g., 5 and 11). If some digits of a number are mistakenly used as exponents and the number's value doesn't change, such a number is called a "printer's error number." An example is 2592, as it equals $2^5 \times 9^2$ = 32 x 81, so if the 5 and the 2 were mistakenly printed as exponents, the result would not change.

XI - Rapid Multiplication from Hell

During the Second World War, Jakow Trachtenberg, a Ukrainian Jewish mathematician who had always been critical of Nazi politics, was imprisoned in a German concentration camp. There, to keep his mind away from the horrors and deprivations of that hell, he "took refuge" continuously in the world of numbers, pondering their unique properties and developing, day by day, an innovative method of rapid mental calculation. The remarkable thing was that he developed this methodology entirely in his mind, never writing anything except to jot down the final results of his work on occasional scraps of paper he managed to find by sheer luck. Fortunately, in 1944, Trachtenberg managed to escape from a prison he had just been transferred to, apparently thanks to the assistance of his wife who, by pawning all her jewelry, bribed some guards and helped him flee. He then escaped to Switzerland, where he ended up teaching his mathematical method, which received several recognitions and was particularly appreciated, especially among those who struggled with traditional calculation methods.

In particular, in this chapter, we will analyze his method of rapid multiplication for 5, 6, 7, 9, 11, and 12, while later we will address his general "graphical multiplication," which has the advantage, just as in the chapter on "instantaneous" column addition, of reducing the number of digits to remember compared to the operation performed in columns using the "classical" method. In fact, Trachtenberg also developed "rapid" multiplication methods for 3, 4, and 8, but they will not be covered here, as I believe they are excessively complex for a book like this, whose aim remains to make mathematics simpler and faster.

I will start by saying that, to some, this method may still seem a bit cumbersome. However, let's still try to consider that:

- Except for some basic operations like "halve" or "double," multiplications are all converted into very simple sequences of subtractions or additions.

- Its speed compared to other techniques increases significantly as the multiplicand (that is, the number you want to multiply by 3, 4, 5, 6, 7, etc.) becomes larger.

- It is a method specifically designed to reduce the number of digits to remember, which, as we know, is ideal for "pure" mental calculation.

- In my opinion, it is an extremely interesting technique from a cultural and anthropological perspective, considering its history, the original approach it offers to arithmetic, and the way it exploits some intriguing properties of whole numbers.

- As mentioned many times before, having multiple tools to solve the same problem is an extraordinary asset from a creative and intellectual standpoint, as it can literally "train" our brain to adopt a mode of creative problem solving that is not necessarily linear and unidirectional.

- Beyond my personal considerations, *everyone is unique,* so it's up to you to compare the various methods at your disposal and understand which one suits you best. For instance, the very fact that multiplications are transformed into additions will be incredibly useful if you still struggle with multiplication tables, which are completely unnecessary here (yes, you can multiply very long numbers by 7 without remembering a single entry from the 7 times table). If you still can't quite "digest" it, you know the advice: forget the Trachtenberg method and move on to the next chapter!

But let's now delve into the "heart" of the Trachtenberg method, and make a few important preliminary remarks on how it works:

- The focus is solely on the multiplicand (the number *to be multiplied*), and operations are performed on it.

- Start with the units digit of the multiplicand and proceed by performing operations one digit at a time, moving to the left. We will later examine in detail how to perform these operations. For now, the important thing is to remember the method and order in which to proceed.

- The digit to the right of another digit is called its *neighbor*. Since every Trachtenberg calculation strategy is based on summing a digit and its neighbor, this is a fundamental concept. Let's take 3792 as an example. The neighbor of 3 is 7, the neighbor of 7 is 9, and the neighbor of 9 is 2. Then, the "neighbor" of the units digit should always be considered as 0.

- The digit on the far left of any number, however, should always be considered as if it were next to an "unseen" 0. This 0, unlike the one we discussed in the previous point, should be treated as if it were indeed a digit of the multiplicand in every respect, and only after operating on it does the process come to a halt.

- A simple method to help you remember how the two zeros work, as discussed in the last two points? At least initially, before proceeding, you could actually write them on either side of the multiplicand, making sure to place the one on the right after a decimal point. For example, you could write 1234 as 01234.0 or 452 as 0452.0. Arranged in this way, the zeros won't arithmetically alter the value of the number and, at the same time, will help you remember much more easily to start from the decimal point and stop at the leftmost zero.

- When the method requires you to "halve" a number, and that number is odd, you must always round down. That is, you should write down the half of the nearest even number below it. So if you're asked to halve 1, write 0; for 3, write 1; for 5, write 2; for 7, write 3; and so on.

- Just as it happened in many techniques illustrated in the previous chapters, each individual sum results in a digit of the outcome. Therefore, if one of them exceeds 9, only the unit is added to the result, while the tens are carried over to the calculation of the next digit.

Let's now see how to implement the actual method:

Multiplication by 11

After the method obtained through decomposition and the Vedic method illustrated in the chapter on golden numbers, here is a third method to multiply by 11: starting from the units and moving to the left, "add each digit to its neighbor." That's it.

Let's immediately look at an example, to better establish what was mentioned in the initial explanation of the method, and try to calculate 56782 x 11:

- If you want, transform 56782 into 056782.0. The value doesn't change, but it will help you remember how to proceed better. We will apply this simplification only in the first example.

- 05678**2**,0
  We start with the 2 in the units place. Its neighbor, as mentioned, is 0. So the first addition to perform is 0 + 2. And here, quite simply, the units digit of the result is indeed 2.

- 0567**8**2,0
  Now let's move to the left and add 8 to its neighbor. So, 8 + 2 = 10. The second digit of the result is 0. Remember that there is a carryover of 1 that you will need to add to the next sum.

- 056**7**82,0
  7 + the neighbor 8 = 15, + the carried over 1 from before gives us 16. The third digit of the result is 6. We will have to carry 1 over to the next sum.

- 05**6**782,0
  6 + the neighbor 7 = 13, + the 1 carried over equals 14. The fourth digit of the result is 4. We will have to carry 1 over to the next sum.
- 0**5**6782,0
  5 + the neighboring 6 = 11, + the 1 carried over results in a 12. The fifth digit of the result is 2. We will need to carry 1 over to the next addition.
- **0**56782,0
  We said that we consider an "non-existent" zero to the left of the 5 as the last digit of the multiplicand. So 0 + its neighbor 5 = 5, plus the carryover from the previous sum equals 6. Thus, the sixth and final digit of the result is 6.
- Final result = 624,602

The sharpest among you have surely already noticed that this method, save for a few details, is practically identical to the Vedic method. Considering that during Trachtenberg's time Vedic Mathematics was neither widespread nor known in the West, it remains, in my view, quite fascinating to observe how people from completely different eras and places arrived at such similar conclusions.

The Exercise

If you'd like to master the Trachtenberg method, try applying the technique you've learned so far to calculate:

898 x 11 = ?

1024 x 11 = ?

34,234 x 11 = ?

Multiplication by 12

In order to multiply by 12, you must, following the "Double each digit and then add the neighbor" method.

So with 11, we simply had to add the neighbor. But with 12, we simply need to double it first.

Curiosity: there is also a specific Vedic Sutra about multiplication by 12 that states "Last and twice the second-to-last," which indicates, with a few details, exactly the same method.

But let's move on to practice and use this technique to multiply 829 x 12:

- 82**9**
  Start with 9. Double it, giving you 18. The neighbor is 0, so 18 + 0 = 18. Write 8 in the result and remember to carry 1 over.

- 8**2**9
  Double 2 to get 4. Add the adjacent 9 to get 13. Add the carried 1 to get 14. Place 4 in the result and carry a 1 over.

- **8**29
  You're at the 8. Double it to get 16. Add the adjacent 2 to get 18. Add that carry-over 1 to get 19. Place 9 in the result and carry the last 1 over.

- **0**829
  Now you are at the "phantom 0" on the far left. Doubling it is still 0. Adding the neighboring 8 makes it 8, and adding the carried over 1 turns it into 9.

- Final result = 9948.

The Exercise

Try applying the method seen so far to calculate:

343 x 12 = ?

9543 x 12 = ?

14,988 x 12 = ?

Multiplication by 6

In order to multiply by 6, you need to "add to each digit half of its neighbor and add 5 if the digit is odd."

The process is similar to x11, except this time you must halve the neighboring number before adding it to the figure you're working on (remembering to round down if the neighboring number is odd, as I mentioned in the introduction). Additionally, if that figure is odd, you need to add 5 to the sum before moving on to work with the next figure.

Let's demonstrate the method by solving 6821 x 6:

- 682<u>1</u>
  Start with the 1 in the units place. The neighbor is 0, so nothing else needs to be added. However, you must add 5 because 1 is odd. Therefore, the first digit of the result is 6.

- 68<u>2</u>1
  To 2, you should add half of its neighbor 1. Half of 1 is 0.5, and rounding it down, 0.5 becomes 0, which doesn't affect the sum. Also, 2 is even, so you don't need to add anything else. The second digit of the result is 2.

- 6<u>8</u>21
  To the 8, you need to add half of its neighbor 2, which is 1, making it 9. Additionally, 8 is even, so you don't need to add anything else. The third digit of the result is indeed 9. Be careful: many confuse this step and mistakenly check if the resulting 9 from the partial sum is even or not. Since 9 is odd, they add 5 and get everything wrong. Try to avoid this mistake and remember to always evaluate the parity of the initial digit of the multiplicand you're working on, and never the one "modified" by subsequent additions.

- <u>6</u>821
  Add half of its neighbor 8 to 6, which is 4. So 6 + 4 = 10. Write

down 0 and carry 1 over. You don't add anything else because 6 was even. The fourth digit of the result is 0.

- **0**6821
You're at the "phantom zero" to the left of the number. Add half of its neighbor 6, which is 3. Plus the 1 you carried over earlier becomes 4. The fifth and final digit of the result is 4.

- Final result = 40,926

The Exercise

Try applying the method to calculate:

134 x 6 = ?

5422 x 6 = ?

98,134 x 6 = ?

Multiplication by 9

Here is the procedure for multiplying by 9:

- For each digit, subtract it from 10 if it's in the units place. Subtract from 9 if it's in any other place. Do not subtract anything if it's the leftmost zero. Perform this subtraction each time on the digit you are dealing with, not on the entire number from the start. This will help you avoid mistakes.

- Add each digit to its neighbor. Be careful: the neighbor always remains the digit originally present in the multiplier before calculating the subtraction indicated in the previous step.

- When you reach the last zero on the left, instead of adding the neighbor, add the neighbor minus 1.

Even here, as you may have noticed, it's possible to catch a glimpse of a "shadow" of a Vedic Mathematics Sutra introduced in the past. Coincidences or patterns that inevitably repeat?

Let's dive straight into an example to dispel any potential doubts. This method might seem a bit more complicated, but in reality, it can be much more straightforward than the "classic" multiplication by 9. Let's try calculating 3825 x 9:

- 382**5**
  (10 - 5) + 0 = 5

- 38**2**5
  (9 - 2) + the neighbor 5 = 12. We place 2 in the result and carry 1 over.

- 3**8**25
  (9 - 8) + the neighboring 2 = 3. Plus the 1 carried over, the third digit of the result is 4.

- **3**825
  (9 - 3) + the neighboring 8 = 14. Place 4 in the result and carry 1 over.

- **0**3825
  0 + the neighboring 3 - 1 = 2 + the carried-over 1 = 3.

- Final result: 34,425.

The Exercise

Try applying the method in order to calculate:

761 x 9 = ?

1323 x 9 = ?

42,345 x 9 = ?

Multiplication by 5

Let's pause for a moment and return to a slightly simpler process than the previous ones. As we know, one initial method for multiplying by 5 is through the *expressions decomposition*, which involves halving a number and then multiplying it by 10, or vice versa. Trachtenberg's method, in fact, doesn't use tools that are all that different, yet it turns out to be much simpler to apply mentally. Namely:

- Halve the neighbor.
- If the number you started to work with is odd, add 5.

Let's immediately demonstrate the method by trying to calculate 24568 x 5:

- 2456**8**
  The neighbor of 8 is 0. Half of 0 is still 0, so the first digit of the result remains 0.

- 245**6**8
  The neighbor of 6 is 8. Halving gives us a 4. The second digit of the result is 4.

- 24**5**68
  The number next to 5 is 6. After halving it = 3. However, 5 is odd, so we must add 5.
  3 + 5 = 8, which is the third digit of the result.

- 2**4**568
  The neighbor of 4 is 5. After halving it (and after rounding down) = 2. The fourth digit of the result is indeed 2.

- **2**4568
  The neighbor of 2 is 4. After halving it = 2. The fifth digit of the result is still 2.

- **0**24568
  The neighbor of the "phantom 0" is 2. Halving it gives us a 1. The sixth digit of the result is precisely 1.

- Final result: 122,840.

The Exercise

Use this method to calculate:

542 x 5 = ?

3941 x 5 = ?

67,134 x 5 = ?

Multiplication by 7

The rule here is: "double each digit, add half of the neighboring digit, and if the digit you started with is odd, add 5."

But let's immediately demonstrate the process by multiplying 2894 x 7:

- 289**4**
  I take the units digit and perform 4 x 2, plus half of 0 which is always 0 = 8, which is the first digit of the result.

- 28**9**4
  Go to the tens: 9 x 2 + 4 / 2 = 18 + 2 = 20. Add 5 because 9 is odd and get 25. 5 is the second digit of the result, and next time we will need to carry 2 over.

- 2**8**94
  Move to the hundreds: 8 x 2 + 9 / 2 = 16 + 4 = 20. Then add the 2 carried over and get 22.
  2 is the third digit of the result, and next time We will need to carry 2 over.

- **2**894
  The thousands: 2 x 2 + 8 / 2 = 4 + 4 = 8. Then add the 2 you just carried over and get 10. 0 is the fourth digit of the result and we will need to carry 1 over for the next one.

- **0**2894
  The Phantom Zero: 0 x 2 + 2 / 2 = 0 + 1 = 1. Add the carryover of 1 and get 2. 2 is the last digit of the result.

- 20,258 is the final result.

The Exercise

Use this method to calculate:

321 x 7 = ?

5123 x 7 = ?

21,321 x 7 = ?

Multiplication by 4

The process for multiplying by 4 is similar to that for multiplying by 8 and by 9 (even though, in my opinion, it may start being a bit too convoluted; still, let's have a look). Specifically, you need to proceed as follows:

- For each digit, subtract it from 10 if it's the unit digit. Subtract from 9 if it's any other digit. Do not subtract anything if it's the last zero on the left.
- Add half of the neighbor and any carry-over.
- Add 5 if the number you started operating from was odd.
- When you reach the last zero on the left, instead of adding half of the neighbor, add half of the neighbor minus 1.

Let's now use this method to calculate 987 x 4:

- 98<u>7</u>
  Units: (10 - 7) = 3. Let's add 5 because the units digit is odd, resulting in 8, which we will place in the result.
- 9<u>8</u>7
  Tens digit: (9 - 8) = 1. Adding half of the neighbor (7 / 2

rounded down is 3) gives us a 4, which we can place in the result.

- **9**87
  Hundreds: (9 - 9) = 0. Adding half of the neighbor, we get 4. However, 9 is odd, so we need to add another 5, resulting in a 9 that we place in the result.

- **0**987
  Last zero: Even after doubling it, it still is 0. We can add half of the neighbor, which rounded is 4. We will then subtract 1 and get 3, which we will place in the result.

- Final result: 3948

The Exercise

Use this method to calculate:

988 x 4 = ?

3124 x 4 = ?

43,181 x 4 = ?

I am fully aware these may not be simple methods to remember. So here's is a little summary:

- **All methods:** proceed from right to left. The starting digit is the unit, and the method stops at the "phantom 0" at the far left of the number. The "neighbor" is considered the number to the right of a digit, and for the units digit, the neighbor is 0. When halving is required, always round down, and when a result is greater than 10, write the unit and carry the tens over.

- **Multiplication by 5:** Halve the neighbor of each digit and add 5 if the digit is odd.

- **Multiplication by 11:** Add each digit to its neighbor.

- **Multiplication by 9:** Subtract 10 from the units digit, 9 from the others, and nothing from the leading zero. Then, add the neighbor to all except the phantom zero, to which you should add the neighbor -1.

- **Multiplication by 12:** Like x11, but double each digit before adding it to its neighbor.

- **Multiply by 6:** Like multiply by 11, but just add half of your neighbor.

- **Multiplication by 7:** Like x11, but double each digit before doing anything else, add half of the adjacent digit instead of the whole adjacent digit, and finally add 5 if the digit in question is odd.

Have you got the feeling that the latest methods complicate your life rather than simplify it? Then discard them if you see fit. But above all, now relax and breathe a sigh of relief, because in the next chapter, I will introduce you to a multiplication strategy so straightforward that you'll probably end up wondering how it could have been completely overlooked in common school curricula.

The Historical Pill

Muhammad ibn Musa al-Khwarizmi was a Persian mathematician, astronomer, and geographer who lived during the Abbasid Caliphate, around 800 AD. He is mainly known for writing the book "The Compendious Book on Calculation by Completion and Balancing," which had a tremendous impact on mathematics. The term "algebra," in fact, derives from the Arabic word "al-jabr," which is one of the two methods used in al-Khwarizmi's book to solve quadratic equations. Al-jabr can be roughly translated as "completion" or "restoration," referring to the classical operation of moving a subtracted term from one side of the equation to the other. This treatise by al-Khwarizmi not only introduced concepts of algebra but also provided examples of how these concepts could be applied to strictly practical matters such as inheritance, property, and measurement. Furthermore, the term "algorithm" derives from the Westernization of his name, reflecting the significant impact he had on mathematics and science as a whole.

The Aphorism

"The beauty of mathematics lies not only in its results but also in its methods. In the way it can transform a complex assumption into an elegant and understandable proof."
(Terence Tao)

Mathematical Oddities

Take an integer. If it's even, divide it by 2. If it's odd, multiply it by 3 and add 1. Repeat the process. The Collatz Conjecture states that, regardless of the initial number, you will eventually reach 1. Despite being "quite intuitive," this conjecture is incredibly difficult to prove and remains one of the open problems in mathematics.

XII - Vertically and Crosswise

This rapid multiplication method for two numbers originates from the third Sutra of Vedic Mathematics, known as "Urdhva-Tiryagbyham," which can be translated into our language as "Vertically and Crosswise." It actually has its precise demonstration even in our polynomial algebra, but we will set it aside and focus on illustrating the method, which is much faster and more convenient than any column multiplication. Let's begin by multiplying two two-digit numbers, for example, 28 x 45:

- Multiply the two tens digits and place the product on the left in the result. Therefore, in this case, we have 2 x 4 = 8____

- Multiply the two unit digits and place the product to the right in the result. So, in this case, we have 5 x 8 = 40 and the partial result becomes 8___40.

- Now multiply the tens of one number by the unit of another and vice versa, then sum the two products together. Therefore, in this case, we have (2 x 5) + (4 x 8) = 10 + 32 = 42. Now we have three partial result numbers: 8__42__40

- Now we just need to *carry over*: as seen in all the addition or multiplication methods illustrated so far, in the final results you must include only the units of the partial results, while the tens are always carried over to the leftmost digit. So, simply put, you have:

The 4 from 42 is carried over to the 8 on the left = 12_2_40

The 4 from 40 is carried over to the 2 on the left = 12_6_0

- When the last two digits are single-digit numbers, the carrying over process is complete, and here is your result: **1260**.

Note that we achieved your impressive 1260 in a quarter of the time it would have taken you by using a more "classic" calculation method, and with the need to memorize far fewer digits.

This method is particularly suitable for "pure" mental math. However, if that's the case and you want to apply it without writing anything, there is an additional tip to consider: start *not from the extremes but from the central numbers*. In other words, begin with what was previously indicated as the third step. By doing so, we can sum "tens x units" and "units x tens" first, which is what most taxes our brain. With our memory completely free, this simplifies the procedure considerably. Let's look at an example of "pure" mental calculation carried out following this method:

- **Problem:** multiply 78 x 44
- **Step one:** multiply the tens of one number by the units of the other and vice versa. Or, if it is visually easier for you, multiply the "inner numbers" together and add the result to the product of the "outer numbers." For example:
- 7**8** x **4**4: the unit of the first number and the ten of the second number are the "inner ones."
- **7**8 x 4**4**: the ten of the first number and the unit of the second number are the "outer ones."
So we can mentally calculate 7 x 4 = 28 and 4 x 8 = 32. We will sum 28 and 32, obtaining 60, and keeping it in mind.
- **Step two:** multiply the two tens. We will calculate 7 x 4 = 28. Place it on the left and keep "28_60" in mind.

- **Step three:** multiply the two units. Calculate 8 x 4 = 32. Place it on the right and keep "28_60_32" in mind.

- We can arrange the carryovers in the order we want, meaning we keep only the units and add the tens to the number on the left. For example, we first carry the 3 from 32, "transforming" the sequence I had memorized into "28_63_2," and then the 6 from 60, finally obtaining "34_3_2." Here is your result = 3432!

Note that if you first calculated "tens x tens" and "units x units" in your head, it would have been much more difficult to simultaneously keep track of the results of these products and the ones to sum in the middle. Try it, and you'll immediately understand what I'm talking about.

We will conclude this chapter by adding that this rapid multiplication procedure comes with an "extension" that allows multiplication of numbers with 3, 4, or 5 digits as well. However, we will address it later because the basic concept is slightly different. Rather than a rapid "mental" multiplication, it involves an ultra-efficient graphical procedure, though still graphical; meaning that it operates according to a precise visual scheme, moving further away from the realm of calculations that can be "easily" done mentally.

The Exercise

Try using the "Vertically and Crosswise" technique to calculate:

12 x 21 = ?

34 x 43 = ?

62 x 89 = ?

73 x 54 = ?

The Historical Pill

While the introduction of negative numbers in Europe occurred during the Renaissance, in China it actually dates back several centuries earlier, specifically to the medieval age. Negative numbers were already used in China in the 3rd century AD, a time when in Europe the idea of numbers "below zero" was considered strange and counterintuitive. It was only around the 13th century, thanks to Arabic translations of ancient Greek and Indian mathematical texts, that Europe began to conceive of these new, "strange quantities." This, of course, was a fundamental turning point in the mathematical and technological progress of the region in question, given its crucial implications in fields such as physics and engineering.

The Aphorism

"The laws of arithmetic are not an invention or creation of the human intellect, but exist independently of it, existing everywhere and at all times. It would be possible to conceive them only as expressions of free thoughts, but not as the creations of those thoughts."
(Carl Friedrich Gauss)

Mathematical Oddities

Staying on the topic of number and quantity "classifications": there is a specific EEC regulation (namely, No. 55 of 11/21/1994) that officially states that multiples of a thousand are called, in order, thousand, million, billion, trillion, quadrillion, quintillion. There do not appear to be, in fact, "official" and "law-abiding" names for larger numbers. And perhaps the most "atypical" aspect of all this is that such a definition was included in a law concerning the transportation of dangerous goods.

Another very large number that might be familiar to many is the "googol," a number consisting of a "1" followed by 100 zeros. The similarity between the name of this number and that of the famous tech giant is not coincidental, as this was originally the name chosen by Google for its search engine. Then (at least according to what the anecdote tells) they decided to change it to "Google" because "Googol" was a name already registered.

XIII - A wonderful Connection

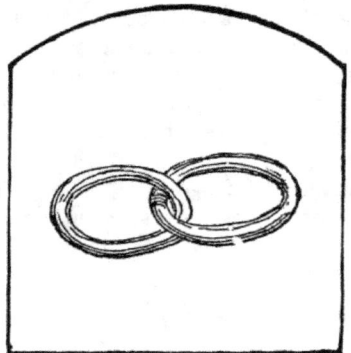

After exploring a "universal" method, we will now return to multiplication techniques that apply "only under specific conditions." The strategies introduced in this chapter allow you to multiply numbers, provided they have "particular" relationships. These techniques are very simple to learn and implement, which means that even though the initial conditions may seem restrictive, they will surely be highly useful whenever you are "fortunate" enough to encounter numbers that meet these criteria. Moreover, as you will see, from these specific cases, it's often possible to easily derive more general ones, thereby further enhancing your arsenal of tools for quick mental calculation.

Two-digit numbers in which either the tens or the units are the same, while the others add up to 10.

This technique is really simple to apply, although it is necessary to distinguish between the two cases in which the procedure varies slightly:

- **Case A:** The tens are the same and the units, when added, total 10. This occurs, for example, when you want to calculate 58 x 52.
- **Case B:** The units are the same, and the tens, when added together, equal 10. This happens, for example, if you want to calculate 23 x 83.

In particular, we have that in "Case A" (for example 58 x 52) we must:

1. Multiply the tens by themselves plus 1 and place the result on the left. (In this case 5 x (5 + 1) = 30. Partial result 30_)
2. Multiply the units together and place the result on the right. (In this case, 2 x 8 = 16. Final result: 3016, achieved in just a few seconds with extremely simple operations)

In "Case B" (for example, 23 x 83), we will instead:

1. Multiply the tens with each other and add the common unit. Then, place the result on the left (In this case, 2 x 8 = 16. Add the sum of the units 16 + 3 = 19. Partial result 19_)
2. As in case "A," multiply the units together and place the result on the right (in this case, 3 x 3 = 9. Final result 1909).

Attention: In both cases, there's a rule to avoid mistakes: if the product of the units results in a single-digit number, it's better to add a 0 in the tens place of the result. Otherwise, for example, in the last demonstration, we might risk writing 199 instead of 1909.

As you can see, these are two extremely simple and straightforward methods. But how can we try to "refine" the restrictions they impose and thus apply them to more "general" multiplications as well?

The answer is in Chapter IX, where we discussed factorizations. That is, we can try to use all the factorization techniques at our disposal to reduce the more general multiplications to those where the condition of this method is respected. Easier said than done, certainly, but let's try to explain it through a specific calculation case; for instance, let's say we need to calculate 58 x 54:

- 54 is very close to 52, a number which, along with 58, would meet the condition of "the same tens and units that, when added, make 10." But what do we do if we have 54 instead of 52?

- We could use the *addends decomposition*. That is, decompose 54 into (52 + 2) and thus obtain:

$$58 \times 54 =$$
$$58 \times (52 + 2) =$$
$$(58 \times 52) + (58 \times 2)$$

58 x 2 is relatively simple, while 58 x 52 can be calculated instantly thanks to this method.

A possible "secret" here, in short, is recognizing when it's possible to break down the operation in such a way that one of the two factors shares the same tens (or units) as the other, while the other digits sum to 10. But more importantly, it involves understanding when it's advantageous to do so. Because, especially in this specific case, there's always the method explained in the previous chapter lurking, which in its simplicity, can easily render this technique redundant. But beyond the repeatedly addressed point that having a more diverse "toolset" at your disposal can only be advantageous, we add that, broadly speaking, this method combined with decomposition can still be beneficial when even one of the following conditions occurs:

- The numbers share the same tens or are still very close to each other.

- The numbers have the same unit

- The numbers are such that their tens digits, when added together, equal 10.

- The numbers are such that their units, when added together, equal 10.

The important thing, therefore, is to "train the eye" so that we can immediately recognize when one of these conditions is met. Clearly, through decomposition, we will then be the ones to divide the addends in a way that allows us to apply our technique most effectively.

Numbers between 11 and 19

This method is very simple to apply and learn. Suppose, for example, you want to multiply 17 x 18. You will need to:

- Add the units of one number to the other. So 17 + 8 = 25.
- Multiply the result by 10, and 25 x 10 = 250.
- Add this last result to the product of the units. Therefore, 7 x 8 = 56, which added to 250 results in a 306, which is our final result.

Unlike the previous method, this one *is limited* and does not offer particular variants, except for the possibility of breaking down multiplicands, by hence transforming "common" multiplications into multiplications between these kind of numbers.

For example, if you need to calculate 12 x 21, you could break down 21 into (19 + 2). Then you would have 12 x (19 + 2) = (12 x 19) + (12 x 2). You can apply this method to the first part, while the other one is immediate.

Similarly, factorization can be easily applied here. Let's for example say you want to calculate 26 x 28. You could factor 26 into 13 x 2 and 28 into 14 x 2. So, you'll have 13 x 2 x 14 x 2. Using the commutative and associative properties of multiplication, you can rearrange it to (14 x 13) x 2 x 2. This means you can first apply this multiplication method to 14 x 13 and then double the result twice.

Numbers that are "equally distant" from any integer

This method involves "replacing" the calculation of a multiplication with the calculation of a square. But what does "equally distant from an integer" precisely mean?

If we imagine all integers placed on a line, then, at least for those among you with a more "natural" inclination for visual imagination, two integers are "equidistant from an integer" if they are both even or both odd, and the integer "in between them" will also be their *arithmetic mean*.

For example, **34** and **36** fall under this definition: the same distance (of **1**) from the number **35**, which is also their mean. The same applies to **80** and **120** (distance **20** from **100**), or **67** and **89** (distance **11** from **78**).

But how can we use all this information to perform our multiplication? Well, the multiplication of two numbers that are "equidistant from an integer" is equivalent to the square of their mean minus the square of their distance (from the mean).

Ok, this wasn't very simple; let's use an example to avoid getting lost. Let's take the numbers listed above: 34 and 36:

- We said their mean is 35. However, their distance from the mean is 1.

- The square of the mean is 1225, and the one of the distance is always 1.

- Square of the mean - square of the distance = 1225 - 1 = 1224, which is exactly the result of 34 x 36!

Another example mentioned above: 67 and 89.

- We have that their mean is 78 and, as stated, their distance from 78 is 11.

- The square of the mean is therefore 6084, and the one of their distance is 121.

- 6084 - 121 = 5963, which is exactly the result of 67 x 89.

It is clear that the "weak point" of this trick lies in the fact that, for it to be effective, it requires us to quickly memorize or learn to calculate beforehand the squares of two-digit numbers. For now, let's not worry too much about this; instead, let's try to understand how this method works in general terms, and later on, in a few chapters, we can learn to approach the "squares" problem with more method, structure, and speed.

Numbers close to a power of 10

This multiplication technique draws heavily from the Vedic Sutras. In particular, let's "dust off" the Sutra introduced in Chapter V, which, through the operation "All from 9 and the last from 10," allowed us to calculate the difference between the power of 10 immediately greater than a number and the number itself. So, when applied to 87, it quickly calculated the result of 100 - 87; when applied to 343, it calculated 1000 - 343; when applied to 7384, it yielded the result of 10,000 - 7384, and so on.

That said, let's first distinguish the three cases in which the technique can be applied:

- **Case 1:** Both numbers are slightly less than a power of 10. For example, 9948 x 9975.

- **Case 2:** Both numbers are slightly greater than a power of 10. For example, 1018 x 1023.

- **Case 3:** The two numbers are such that one is slightly below and the other slightly above. For example, 883 x 1007.

Let's begin by illustrating case 1, which is the simplest; we will soon see, moreover, that the others are merely simple variations of it:

1. Keep in mind that if both numbers are sufficiently close to the next power of 10, the result will have a number of digits equal to the sum of the number of digits of the operands. For example, the result of 998 x 999 will have six digits, 9876 x 9889 will have eight, and so on.

2. Calculate the complement of both numbers to the nearest higher power of 10. If it's not straightforward, apply the Sutra "All from 9, the last from 10."
3. Multiply these complements and place it on the right in the result.
4. Subtract the complement of one number from another and place it on the left in the result (here you can choose the one that seems easiest to you, as the outcome will be the same).
5. If the number of digits in the partial result is less than that calculated in the first step, add "zeros" in between.

Let's immediately look at an example by trying to calculate 9948 x 9975:

The two numbers have 4 digits, so the result will have 8 digits.

The complement of 9948 concerning 10,000 is 52.

The complement of 9975 concerning 10,000 is 25.

Multiply 52 by 25, obtaining 1300. These will be the four digits on the right in the result.

Calculate 9975 - 52 or 9948 - 25 (same result), obtaining 9923. These will be the four digits on the left in the result.

Here's the final result: 99,231,300, achieved in just a few seconds and without any effort!

Note that the extraordinary efficiency of the method lies in the fact that it transforms a 4-digit multiplication into a multiplication between 2-digit numbers. The closer the two factors are to a power of 10, the simpler the method becomes. For example, 9997 x 9998 reduces to a simple multiplication of 2 x 3. We could in theory use this method with every integer number but, obviously, calculating complements far from a power of 10 involves multiplications exceeding 3 or 4 digits, hence diminishing its efficiency.

Let's move on immediately to case 2, which is very similar to the previous one:

1. Keep in mind that if both numbers are sufficiently close to the lower power of 10, the result will have a number of digits equal to the sum of the number of digits of the operands minus 1. For example, the result of 1002 x 1011 will have 7 digits.
2. Calculate the complement of the operands to the power of 10 immediately below. Here, clearly, the calculation is immediate.
3. Multiply the results and place it on the right in the outcome.
4. Add to a number the complement of the other and place it to the left in the result.
5. If the number of digits is greater than what was calculated in the first step, you will need to take the leftmost digit from the right section and add it to the left. For example, if you have 1345_4393, and you needed the result to be 7 digits, you should take the 4 and add it to 1345, thus obtaining 1,349,393.

Let's immediately do an example by calculating 1072 x 1048:

Each factor has 4 digits. 4 + 4 - 1 = the result must have 7 digits.

The complements to the power of 10 immediately below are obviously 72 and 48.

Multiply them, obtaining 3456. Place it on the right.

Calculate 1072 + 48, or 1048 + 72, in order to get 1120. Place it on the left.

The final result is 1120_3456. Note that the total number of digits exceeds what the result should have. So we shift the excess digit on the right towards the left. In this case, the excess digit is a 3, so we get = 1123_456. And, in fact, 1,123,456 is the final result.

Finally, let's examine case 3, where one number is slightly larger than a power of 10 and the other is slightly smaller:

1. Add to the smaller number the complement of the other number to the considered power of 10, or subtract from the larger number the complement of the other number to the considered power of 10. Both operations will yield the same result.
2. Append zeros corresponding to the number of digits in the considered power of 10..
3. Multiply the two complements together and subtract them from the number obtained in the previous step.

Let's demonstrate the procedure of the method by trying to multiply 989 x 1024.

- The two complements are 11 (989 + 11 = 1000) and 24. We can either add 24 to 989 or subtract 11 from 1024; we would get the same number. We choose to subtract 11 from 1024 = 1013.
- The power of 10 considered is 1000, so we place three trailing zeros: 1,013,000.
- 11 x 24 = 264 (and you remember how to multiply by 11, right?)
- We need to subtract 264 from 1,013,000. We can speed that up with a simple trick seen in Chapter V: 1,013,000 = 1,012,000 + 1,000. So, we could first calculate 1,000 - 264 by applying the already seen Sutra for these cases, and then add 1,012,000. This yields 1,012,736, which is also the result of 989 x 1024.

Since this last case might be a bit more "challenging" for some, let's do one final example by trying to calculate 973 x 1003.

The two complements are 27 and 3. By adding 3 to 973 (or subtracting 27 from 1003), you will get 976.

The power considered is always 1000, so add three zeros: 976,000.

Multiply 27 by 3, obtaining 81.

Subtract 81 from 976,000, obtaining 975,919, which is precisely the result of the initial multiplication.

A summary and some exercises.

Since by now you've probably got "flooded" with information, let's do a quick recap for those who, before moving forward, need to organize and consolidate thoughts a bit:

When the tens are the same and the units add up to 10: Multiply the tens by themselves + 1 and place the result on the left. Then, multiply the units together and place the result on the right. To further reinforce this concept, try using it to mentally calculate:

88 x 82 = ?
43 x 47 = ?

When the units are the same and the tens add up to 10: Multiply the tens together, add the common unit, and place it on the left. Then, multiply the units and place the result on the right. Remember to add zeros where necessary. To consolidate this principle further, try using it to calculate:

21 x 29 = ?
78 x 72 = ?

When the numbers range between 11 and 19: Add the units of one number to the other number. Multiply the result by 10. Add to this the product of the units. Use this to calculate:

11 x 18 = ?
13 x 17 = ?

When the numbers are equidistant from an integer (both even or both odd): Find the square of their average, minus the square of their distance from the average (yes, not always the simplest). Use this method to calculate:

28 x 30 = ?
13 x 19 = ?

When numbers are slightly less than a power of 10: Determine the number of digits in the result. Multiply the complements of both numbers and place it on the right. Subtract one number's complement from the other and place the result on the left. Add zeros in between if necessary. Use this to calculate:

9912 x 9998 = ?
9930 x 9974 = ?

When numbers are slightly more than a power of 10: Determine the number of digits in the result. Multiply the complements of both numbers and place it on the right. Add one number's complement to the other and place it on the left. Make any necessary carry adjustments. Use this to calculate:

1131 x 1008 = ?
1047 x 1091 = ?

When one number is slightly less and the other slightly more than a power of 10: Find the complements of both numbers with respect to the nearest power of 10. Subtract one number's complement from the other or add one number's complement to the other. Append as many zeros as there are in the power of 10. Multiply the two complements together and subtract the result from the previous number. Use this method to calculate:

989 x 1024 = ?
973 x 1003 = ?

The Historical Pill

The Cartesian coordinate system, a fundamental tool of modern mathematics, is attributed to the French philosopher and mathematician René Descartes (1596-1650). The legend has it that Descartes, while bedridden with illness, watched a bug move across the ceiling; thus, he envisioned a system in which he could describe the insect's position using two numbers: the insect's distance from two perpendicular walls of the room. This concept, perhaps very simple today but revolutionary at the time, laid the groundwork for analytic geometry and helped forge a connection between algebra and geometry, completely transforming the way we approach mathematical problems.

The Aphorism

"For those who do not know mathematics, it is difficult to feel the real beauty, the deep beauty of nature... If you want to know nature, to appreciate it, you must understand the language it speaks."
(R. Feynman)

Mathematical Oddities

Geometric series are sequences of numbers in which each subsequent term is the product of the previous term and a constant. For example, 1, 2, 4, 8, 16 is a geometric series where the constant is, of course, 2.

The "oddity" of these series is that if you add up ALL the terms of an infinite geometric series, you might end up with a finite number. For example, if you start with 1 and add half of the previous number, continuing infinitely (1 + 1/2 + 1/4 + 1/8 + ...), mathematics tells us that you will not get infinity but, simply, 2. This mathematical result may seem extremely counterintuitive, but it forms the basis of many physical, economic, and computer theories; for example, in telecommunications engineering, geometric series help calculate the total data that can be transmitted through a network.

XIV - Chinese Graphical Multiplication

The first of the graphical multiplications that we will introduce in this book, the so-called "Chinese Graphical Multiplication," always exerts a certain charm on those who discover it for the first time. This is partly due to its extraordinary simplicity, partly because of its genuine speed compared to column multiplication (which so far, as we have seen, seems to be the most inefficient procedure of all), and partly because, in effect, it allows you to operate by simply drawing "sticks" without even knowing a multiplication table. Furthermore, if you don't have paper and pen with you, it's even possible to put it into practice by suitably arranging sticks, matches, raw spaghetti, toothpicks, or any set of stick-shaped objects available.

But here's how this "curious" calculation technique works:

- For each digit of the first number to multiply, you should draw, from bottom to top, a corresponding number of groups of diagonal lines oriented like the "\" symbol (so, for example, two digits = two groups of lines, three digits = three groups of lines, etc.). Each group must also have a number of lines equal to the value of the digit. For instance, if the number to multiply is "12," you should first represent it by drawing one line and then two more, arranged as follows:

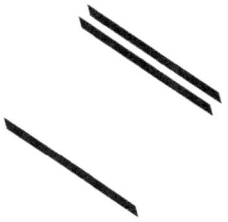

- For each digit of the second number to be multiplied, you should draw, now from top to bottom, an equal number of groups of lines, but oriented like the symbol "/." For example, if you have "31," draw these lines, overlapping them with the previous ones, for the 12, in this way:

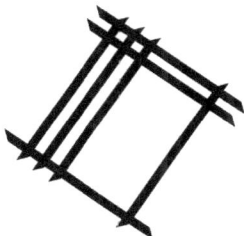

- In certain parts of the drawing, you will notice that the lines belonging to the two numbers have crossed, and indeed, it is possible to group the nearby crossings together. Circle these groups if it makes it easier for you and count how many crossings are present in each group.

(obviously here you can choose whether to actually count the intersections of the group or to multiply the number of lines entering that group horizontally and vertically. Do whatever is most convenient and immediate for you.)

- You will notice that some of the groups of intersections are aligned vertically, as if they were "in column." Then, add the numbers of the intersections of the groups that are aligned this way and write the total below. Here, for example, the two central groups are aligned, and when summed, they equal 7. Therefore, write the number 7 below. Ignore the numbers of the groups that "stand alone."

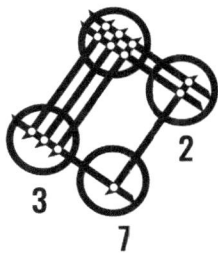

- Look at the figures obtained, and here is the result of your 12 x 31 = 372.

The more "astute" among you may be wondering at this point what would have happened if the central addition had resulted in a number greater than ten. However, those who are "truly astute" have probably already answered: the units digit would have been retained, and the tens digit would have been carried over to the left.

But since the best way to firmly imprint a process is through multiple demonstrations, let's immediately see another example, this time with two three-digit numbers, and try to multiply, for instance, 123 x 311:

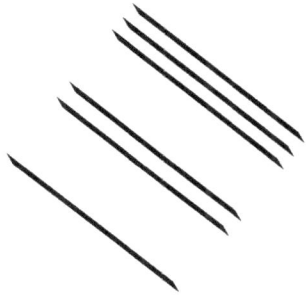

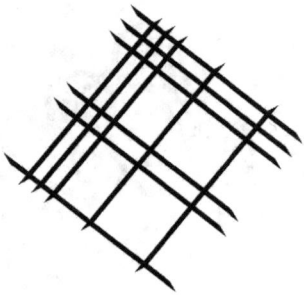

As you can easily notice, the representation of 123 is formed by groups of one, two, and then three sticks in the \ direction, while the 311 is formed by groups of three, one, and the last stick in the / direction. Now let's highlight the groups of intersections and count the intersections for each group.

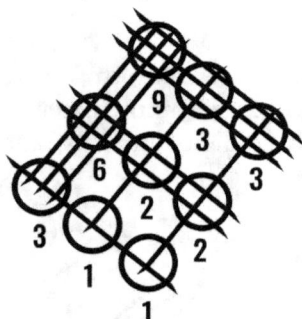

Afterward, we sum down the numbers of the intersections related to the vertically aligned groups:

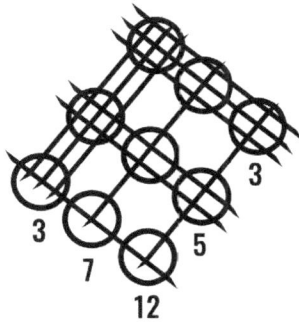

So, by performing the only necessary carryover, we obtain 38,253, which is indeed the final result.

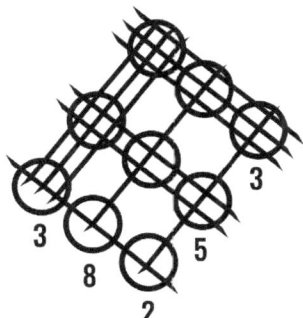

The method can easily be used for numbers with 4, 5, or more digits. Clearly, as the number of digits increases, as well as the values of those digits, the number of intersections to count will also increase, consequently raising the complexity and length of the resulting graphical method. However, despite all this, it remains a method that I would not hesitate to call "quite efficient," easy to use, and with undeniable "recreational" value, especially for those who wish to experiment with different calculation strategies from those we use every day.

Exercises to strengthen your technique. Try using the "graphic multiplication" technique, in whatever practical implementation you prefer (use sticks, spaghetti, Shangai sticks, toothpicks, etc.), to calculate:

24 x 12 = ?

24 x 43 = ?

141 x 222 = ?

The Historical Pill

Blaise Pascal (1623 - 1662) is known for his contributions to projective geometry, probability theory, and even for inventing one of the first mechanical calculators, the so-called "Pascaline." He is also particularly known for his "wager." During a period of religious reflection, Pascal developed an argument stating that, mathematically speaking, it made more sense to believe in God than not to believe. His reasoning was: if you believe and God exists, you gain heaven, whereas if you believe and God does not exist, you lose nothing. Conversely, if you choose not to believe in God and are "wrong," you risk hell. Regardless of whether one agrees with its foundations, Pascal's wager is interesting both for its historical value, reflecting a time when it was not unusual for a mathematician to engage with metaphysical issues, and for the very fact that the reasoning behind it can be seen as a precursor to "game theory." Using a more contemporary language, one could say that, according to Pascal, believing in God was a "dominant strategy."

The Aphorism

"Mathematics is a discipline where the effort required to identify the problem is 90% of the solution. In times of crisis, imagination is more important than knowledge."
(Andrew Wiles)

Mathematical Oddities

The mathematician David Hilbert (1862-1943) proposed a thought experiment known as Hilbert's Hotel. Imagine a hotel with an infinite number of rooms, all of which are occupied. However, a new guest suddenly arrives wanting a room. To accommodate them, you simply ask each guest to move to a room with a number one higher than their current room (the guest in room 1 moves to room 2, the guest in room 2, and so on); since the number of rooms is infinite, this will always be possible. However, this contradicts the initial hypothesis that the hotel was originally full and highlights some of the peculiarities that emerge when we begin to treat infinity as if it were just an "ordinary number."

XV - The Sliding Cross

Returning to more "classic" multiplication methods, let's introduce the "general multiplication" system devised by Trachtenberg that we mentioned a few chapters ago. This system is very simple and effective, even with very long operands. Unlike the "Multiplication Table" devised by the mathematician, it does not require memorizing lengthy or complicated procedures. It only requires a bit of "visual" training to understand the simple "cross" graphic scheme through which operations are performed each time.

To simplify things for you, I will explain only how to multiply two numbers with the same number of digits. If you need to multiply numbers with a different number of digits, you just have to add zeros to the left of the smaller number until it matches the digits of the larger one (which will obviously cause several "cancellations of the product" and will significantly simplify some steps in the method).

Having said that, let's begin by illustrating how to proceed with multiplying two-digit numbers.

Step 1:
Align the numbers in columns and multiply the units together. The result of this multiplication should be placed on the right in the partial result.

Step 2:
Multiply the units by the tens and the tens by the units. Then add the products together. The result of the sum should be placed in the partial result, to the left of the result from the previous step.

Step 3:
Multiply the tens together. Place the result of this product to the left of the result from the previous step, forming a partial result.

Step 4:
If any result from the previous steps exceeds 10, the tens place should be carried to the left.

Yes, this is none other than the multiplication method introduced in Chapter XII. I thought of presenting it again just as it is because, as previously mentioned, the underlying scheme is the basis for multiplying larger numbers, so it's good to train your eye to work in a "cross" manner. But let's immediately see an example of applying the method by multiplying 54 x 78:

54 x
78 =

Step 1:
Align the numbers and multiply 4 by 8, obtaining 32. Place it on the right in the partial result.

Step 2:
Multiply 5 x 8 and add the result of 4 x 7 = 40 + 28 = 68. Place it to the left of the previous result. The partial result will be 68_32.

Step 3:
Multiply 7 x 5 and place it on the left in the result. You will get 35_68_32.

Step 4:
Bringing the 6 over the 35 and the 3 over the 8, you will have 4212, which is precisely the result of the multiplication.

But since this is a method we've already seen, let's move on to the "next level" right away and tackle a multiplication of 3 digits by 3 digits immediately:

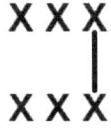

Step 1:
Multiply the units together and place the result on the right.

Step 2:
Multiply the units by the tens and the tens by the units. Add the two products together, and place the result of the sum to the left of the result from the previous step.

Step 3:
Here is the step never attempted before, even though the principle is always the same: multiply the numbers symmetrically "opposite" to each other and then add the results of the products. So calculate (Hundreds x units) + (Tens x tens) + (Units x hundreds). Finally, place the result to the left of the result from the previous step.

Step 4:
Now you are shifting the "cross" to the left and the method is nearly complete. Multiply the hundreds by the tens and the tens by the hundreds. Then add the products and place the result to the left of the result from the previous step.

Step 5:
Multiply the hundreds together and place the result of the product to the left of the result from the previous step.

Step 6:
Carry over.

In practice, what this method requires, regardless of the number of digits, is that:

- Always start from the right side of the two numbers.
- You always start by multiplying **Units x units.**
- One digit at a time, moving to the left, consider progressively larger "pieces" of the two numbers.
- For each "piece" considered, multiply each digit by the one visually opposite to it, sum the results of these products, and place the result of the sum in the partial count.
- When this "piece" is expanded to include the whole number, increasingly smaller pieces of the two numbers are considered this time, obtained by "cutting" the right parts one digit at a time. Thus, the same procedure of multiplying opposing numbers and summing the products is repeated.
- You stop when the "cross" has become a "line" and you've reached **"Leftmost digit x Leftmost digit."**
- If the two numbers happen to have a different number of digits, proceed as if the shorter number had leading zeros to equalize them (for example, 1778 x 32 can be "transformed" into 1778 x 0032).
- Apply all the necessary carryovers.

This, I repeat, is not intended to be a formal or rigorous description of the method, but merely a brief overview of the visual framework to adopt. Once you've learned that, you can apply it to factors of any length.

Let's immediately go through a practical example of the 3x3 method by trying to multiply 673 x 231:

Step 1:
Multiply the units 3 x 1 = 3, place that on the right in the final result.

Step 2:
Multiply the units by the tens 7 x 1 = 7 and the tens by the units 3 x 3 = 9.
7 + 9 = 16, which you can place to the left of the previous result. Now the partial result is 16_3.

Step 3:
Multiply the units by the hundreds 6 x 1 = 6, the tens by the tens 7 x 3 = 21, and the hundreds by the units 1 x 6 = 6. 6 + 21 + 6 = 33. Place that on the left in the partial result, which is now 33_16_3.

Step 4:
Multiply the hundreds by the tens, 6 x 3 = 18, and the tens by the hundreds, 2 x 7 = 14. 18 + 14 = 32, which you can place to the left in the partial result, making it 32_33_16_3.

Step 5:
Multiply the hundreds together: 6 x 2 = 12. The partial result now becomes 12_32_33_16_3.

Step 6:
Calculate all carryovers, resulting in 155,463.

We conclude this chapter by showing how to calculate multiplication of 4-digit by 4-digit and 5-digit by 5-digit numbers. Since the pattern simply repeats itself, and I believe it's much more useful to help you "visually internalize" it rather than endlessly repeating the same concepts, this time I will leave you with images of the process, without explaining it step-by-step any longer.

Multiplication of 4-digit by 4-digit numbers

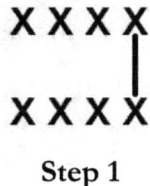

Step 1

Step 2

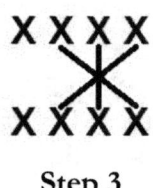

Step 3

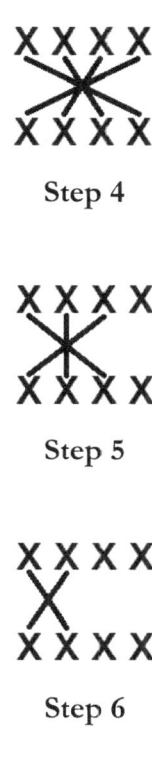

Step 4

Step 5

Step 6

Step 7

Step 8: calculate carryovers

Multiplication of 5-digit by 5-digit

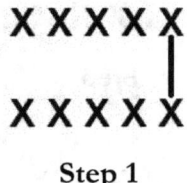

Step 1

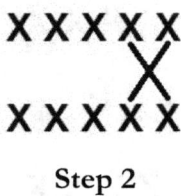

Step 2

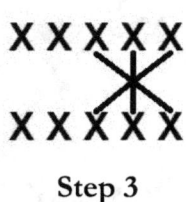

Step 3

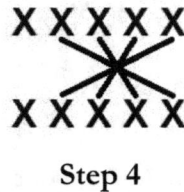

Step 4

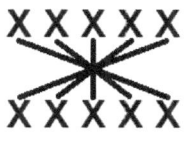

Step 5

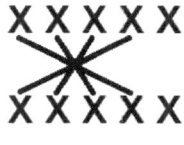

Step 6

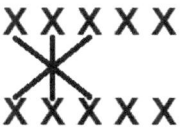

Step 7

Step 8

Step 9

Step 10: calculate the carryovers

A tip for learning to mentally apply this technique might be to use some memory techniques, such as "phonetic conversion" seen in Chapter III, to more easily remember partial results and carryovers. With some practice, you will not only be able to calculate long multiplications in your head, but you will also learn to do it increasingly faster.

The techniques of rapid multiplication are thus concluded with this chapter. However, in the following chapter, we will practically apply them by tackling two topics that are both a "nemesis" for many and absolutely essential in daily practice: percentages and calculations with decimal numbers.

The Exercise

Try using the "sliding cross" technique to calculate:

738 x 291 = ?

506 x 429 = ?

1832 x 4769 = ?

9017 x 6254 = ?

27394 x 58561 = ?

84975 x 62340 = ?

The Historical Pill

John Napier (1550-1617) was a Scottish mathematician primarily known for publishing a work in 1614 titled "Mirifici Logarithmorum Canonis Descriptio," in which he introduced the concept of logarithms for the first time. This mathematical innovation was crucial for simplifying complex calculations, especially in astronomy, a field that was undergoing a revolution at the time thanks to the work of astronomers like Copernicus, Galileo, and Kepler. Before the invention of logarithms, multiplying and dividing large numbers were time-consuming and prone to errors. However, with logarithms, Napier provided a tool that converted these operations into simple additions and subtractions, and in an era without calculators, this simplification of calculations laid the groundwork for significant scientific and technological advancement. He also invented what is known as "Napier's bones," a manual device used for multiplication, division, and extracting square roots.

The Aphorism

"Mathematics too is a science created by human beings, and therefore, in every era and among every people, it has its own unique spirit."
(Hermann Hankel)

Mathematical Oddities

Jorge Luis Borges (1899 - 1986) was an Argentine writer who, like few others, managed to blend in his stories the warmth of human passion, the mystery of myth, and the bewilderment created by mathematical paradoxes. In particular, in his "Library of Babel," he tells of a paradoxical infinite library, composed of an infinite number of books, with all possible combinations of letters. So, if this "impossible" library were ever created, could there exist a book that narrates, in detail and with absolute precision, every moment of your life, from your first breath to your last? And one that tells the same story, but "changes the ending"?

XVI - Travel, Discounts, and Elections

Knowing how to work with percentages is fundamental. A percentage represents the "parts" of something and answers the question "if this thing were made up of a hundred parts, how many parts am I considering?" This makes it, broadly speaking, intuitive for most people (almost everyone can easily visualize 50% or 25% of a cake, for example). This is why the concept is practically everywhere in everyday language: just think of opinion polls, simple store discounts, or the theory of probability that we will tackle in a few pages. Yet, despite being considered an "intuitive" representation, the situation is quite different when it comes to actually calculating and reasoning with percentages, a procedure for which our brain can be easily deceived. An example can be found in the problem: "If the price of a liter of gasoline is 100, and after one week it decreases by 10% and then the next week it increases by 10%, how much will the liter of gasoline cost in the end?" Well, unless one already knows the answer or has previously worked on these types of problems, it's very likely that most people would answer "100," forgetting that the percentage must always be recalculated based on the current value. In fact, 100 - 10% = 90. Then, 90 + 10% = 99, which is the correct answer to the initial question. Or, just think about how often the media "play" with people's difficulty in reasoning concretely with percentages; a very simple example of this can be found in the words of an American reporter who, following the 2008 presidential elections, stated the following:

In 2004, 37% of voters identified as Republican, 37% as Democratic, and the rest as independent. Four years later, 39% of voters identified as Democratic and 32% as Republican; therefore, within four years, 5% of Republicans became Democratic.

Quiz: Could you identify right away where the problem lies in the conclusion of this reasoning, in terms of understanding reality? Take a couple of minutes to think about it, and then read the solution just below.

...

...

...

...

...

Solution: The only certain truth we can derive from the 2008 data is that the Democrats gained a +5% of eligible voters, and not from the Republican electorate; all with the added consideration that, since the total number of voters changed from 2004 to 2008, the statement becomes even more abstract and imprecise.

From all this, we could therefore try to deduce two fundamental truths:

- *A mathematical truth who advises always being careful to calculate percentages based on the actual and "current" value of something.*
- *A "less mathematical" one that advises never to fully trust the percentages thrown around by any politician or biased journalist.*

Having made these necessary preliminaries, let's introduce some more practical tools for their calculation. First of all, as mentioned before, considering the A% of a quantity means dividing that quantity by 100, multiplying it by "A," and considering the result. This can clearly lead to simplifications that, by exploiting certain properties of fractions and the quick multiplication table shown in Chapter IX, allow for a very quick calculation of various percentage values. Therefore, here is a practical table for calculating the most common percentages, created precisely by using these criteria. Naturally, do not aim to memorize it but rather try to "discover" which criteria were used each time, so that you fully grasp the underlying reasoning

2% = Divide the number by 100 and then double it (or vice versa).

5% = Divide the number by 10 and then halve it (or vice versa).

10% = Divide the number by 10.

15% = Triple the number, then halve it and divide by 10 (in any order you prefer).

20% = Divide by 10 and double (or vice versa).

25% = Halve the number twice.

30% = Triple and divide by 10. If having an exact result isn't crucial, you could also approximate the calculation of 30% to 33.3% by dividing directly by 3.

40% = Divide by 10 and double twice (or vice versa).

50% = Halve it.

60% = Double it, then triple it, and finally divide by 10 (in any order you choose).

70% = There are no tricks beyond multiplying by 7 and then dividing by 10. If it's not crucial to have a highly accurate result, you could also approximate the calculation of 70% to 66.6% by doubling and then dividing by 3.

75% = Triple it and then halve it twice (or vice versa).

80% = Double it three times and then divide by 10 (or vice versa).

90% = Triple it twice and then divide by 10 (or vice versa).

The exercise

- How much would a 60% discount on a car costing 2,500 dollars be equivalent to?

- How much would a 15% discount on a bathtub costing 1,955 dollars be equivalent to?

The exercises, as in every other section of the book, are certainly not "mandatory," and are more to be considered as a "bonus" for all well-intentioned readers who want to try to get the most and the best out of this reading rather than an integral part of the book's "core." However, one thing that might be worth emphasizing in this context is that (spoiler!), the second of the two exercises just discussed results in a fractional number, meaning with a decimal point. In fact, unless we are dealing with the rare case of always operating with whole numbers or approximations to whole numbers, the discussion of percentages is necessarily tied to numbers of this kind, and because of this, they will have their dedicated discussion in the following section:

Rounding, nihilism, and "catastrophic" errors

A primary, fundamental point worth emphasizing when discussing decimal numbers is purely pragmatic in nature. It echoes what was just mentioned about approximations and concerns the fact that working with fractional numbers only makes sense in light of the degree of precision we can afford and what is necessary in a given context. But to better understand what I'm talking about: how would you answer this question?

Is 1.73 equal to 1.72999999999999?

Clearly, from a purely mathematical standpoint, the answer is no, nada, nein. But if we were measuring a person's height and those values were in meters, the questions we should be asking here revolve around how crucial it really is to make this distinction and whether we would actually be able to perceive or detect it using any typical measurement tool. For instance, even if we had a precise laser meter capable of distinguishing at the nanometer level, the person's height might vary slightly due to changes in posture throughout the day or hydration levels. Thus, for any conceivable practical purpose, it might be perfectly fine to approximate both heights to a more pragmatic and manageable 1.73m, or 173 centimeters, effectively equating the values mentioned earlier. This naturally stems from the principle that the very physical reality we live in, along with our need to work with manageable quantities, necessitates that we always work with approximated figures; the key is to understand to what degree they are approximated and whether this approximation is "acceptable" for the case at hand. When you think about it, this gives us an interesting, and perhaps "unsettling" fact about our existence: we probably will never know, nor can we ever truly know, how tall we are, how much we weigh, or the exact value of any other measurement within our physical world. It all, and always will be, an approximation.

All this reasoning, however, beyond the sense of "existential bewilderment" with which I tried to provoke you playfully, converges into a truth, for which both humans and electronic calculators are extremely grateful. This truth is that you can always try to speed up calculations with decimal numbers through approximation. Typically, this approximation is achieved through what is called rounding: to round any number, you simply select a part of the fractional number given (for example, the second digit after the decimal point) which is considered "irrelevant" for the case or purpose at hand, and eliminate it. After that, if the leftmost digit among those you eliminated was greater than or equal to 5, you increase by 1 the rightmost digit "remaining" in the number. So, for example, if you have 6.489 and intend to eliminate the last two digits, you would round it to 6.5, since the leftmost eliminated digit (the 8) was greater than 5. Or, if you wanted to round the same number to the whole number, it would simply become a 6 (and it would have become a 7 if it had been 6.589 or 6.689, and so on).

The rounding method calculated in this way, among other things, is not the only technique used in real cases to strip a fractional number of irrelevant digits. Often, in fact, there may be instances where you use what's known as rounding up (always increasing the rightmost, or least significant, remaining digit by 1; thus both 6.4 and 6.5 would be rounded up to 7), or rounding down, also called "truncating" (where you remove the unimportant digits without making any other adjustments; thus 6.4 and 6.5 would be truncated to 6). For anyone wondering which technique is best to apply, the answer will obviously depend on the specific context and the needs we are dealing with. Take, for example, a civil engineer designing a bridge, or an aerospace engineer developing a rocket project. It's clear that, for safety and accuracy reasons, there is a need for extremely high precision in calculations; however, it might also happen that rounding up is preferable if it ensures that estimations are always on the side of "adequate caution" required in such cases. Barring critical or technical scenarios, the "classic" rounding remains the most universally valid technique, as well as the most adaptable to the various contexts we realistically encounter.

Advising engineers, or potential ones, to refer to other manuals for understanding how to build a bridge or an airplane, we continue this section with a few considerations more closely related to our domain, as they are more tightly connected to rapid calculation procedures. If, for example, we need to add or subtract decimal numbers, the process is straightforward: you figure out if and to what extent it's necessary to remove some significant figures, align them accordingly, and apply all the "normal" addition and subtraction techniques discussed in previous chapters.

As for multiplications and divisions, however, let's make a few additions, first of all noting that rounding the operands of these operations can lead to a phenomenon called "error amplification" or approximation: consider, for example, a simple experiment where you need to multiply a number by itself, say, ten times. Let's start with a precise number: 1.01. Multiplying 1.01 by itself ten times, you'll get approximately 1.1046.

Now, however, let's consider the same experiment, but by approximating 1.01 to 1.0. Multiplying 1.0 by itself 10 times will, of course, still yield 1. Notice what happened? An initial approximation of just about one hundredth has caused an approximation, or error in the final operation, of one tenth, which is ten times greater; this might even be "still acceptable" in the practical cases we might encounter. However, it should encourage us to pay further attention when assessing our "degree of acceptable approximation," especially in situations involving continuous iterations of multiplications and divisions; these particular cases could easily lead to errors much larger than what we initially defined as "manageable."

Not without reason, this is a crucial factor for all designers of high-criticality components, such as those mentioned earlier. This is especially true in contexts where these calculations go through a computing tool that does not offer the necessary precision. Given that the capacity of some software or hardware to effectively represent numbers with many digits (and thus with the required degree of precision) is limited, it becomes necessary to pay attention to all cases where repeated operations can significantly "skew" the final results, thus, unfortunately, leading to exploded spacecraft and collapsed bridges.

Having made all the "necessary" philosophical-mathematical premises, and set aside the trivial case of addition, let's start from the next section to better understand how we can multiply and divide in the fastest and most efficient way possible with our fractional numbers.

"Killing" the decimal point… and expressions decomposition!

A more "classic" method to multiply decimal numbers (which, for the sake of simplicity, we will consider as already approximated to our "ideal" degree of precision here) is to treat them as if they were whole numbers, temporarily ignoring the decimal point, and then "reapplying" it at the end.

More specifically: first remove the decimal point and multiply the numbers as you would normally. After obtaining the result of the multiplication, however, it will be necessary to reintroduce the decimal point, and this will be done based on the total number of decimal places present before the removal. For example, in the case of 8.13 x 5.20, we have two decimal places plus two decimal places in the operands, so the result should have four decimal places; in the case of 12.123 x 3.22, it should have five; in the case of 9.134134 x 2.1, there should be seven. And so on.

But let's also quickly look at a complete example: if you find yourself needing to multiply 7.76 x 9.31, you can first multiply 776 x 931 (and here you might use the technique of "numbers close to a power of 10"), obtaining 722,456. Then, since both numbers had 2 digits after the decimal point, you will place it on 722,456 in such a way that after it there are 2 + 2 = 4 digits. So in the end, you will have 72.2456.

With division, on the other hand, in order to try to reduce the operation to one involving integers, it would be advantageous to use the invariant property (that is: use the fact that multiplying or dividing both the dividend and divisor by the same quantity does not change the result). You can do this by multiplying both operands by a power of 10 to eliminate the decimal point.

So, for example, if you have 1.24 / 2.33, you can multiply both numbers by 100 and calculate 124 / 233. Similarly, if you have 133 / 0.002, you can multiply both numbers by 1000 and you'll find that the result is equivalent to 133,000 / 2.

However, one can also try to do even better: a technique that could prove to be much more efficient than the one we have seen so far, in fact, is the one that refers to *expressions decomposition* examined in Chapter IX. In particular, we will have that:
- We can factor the decimal number we're interested in as if it didn't have a decimal point. For example, we can "transform"

our 4.5 into 45, which, when factored, would be equal to 3 times 15 = 3 x 3 x 5.

- "Transforming" 4.5 into a 45 in the previous step means at the end we need to divide by 10 in order to "compensate" for the previous multiplication. So, quite simply, if 45 = 3 x 3 x 5, and 45 = 4.5 / 10, then 4.5 = (3 x 3 x 5) / 10.

The Exercise

Your car consumes an average of 6.6 liters of fuel every 100 kilometers. You are planning a trip of 990 kilometers. How many liters of fuel do you need to complete your trip? (and yes, try to use the *expressions decomposition*).

Addends decomposition

Another method you can use for multiplication with decimal numbers is to break them down into addends. This method, on one hand, has the usual limitation of not being applicable to division, but on the other hand, has the great advantage of being usable regardless of the ability to break down into factors (which, as we should know well, ceases to be useful as soon as we encounter a prime number). But let's illustrate it right away and let's say, for example, we want to calculate 5 x 1.44.

1.44 can be broken down into (1 + 0.4 + 0.04)

As we should know by now, any decimal number is equivalent to the whole number obtained by removing the decimal point, divided by the power of 10 with as many zeros as there were digits after the decimal point that was previously removed. Simply put, for anyone who might be confused: 1.2 is equivalent to 12 / 10; 1.567 to 1567 / 1000, and so on. So, for instance, in this case:

$$0.4 = 4 / 10$$
$$0.04 = 4 / 100$$

In this case, therefore, you could "simply" calculate:

$$5 + (20 / 10 = 2) + (20 / 100 = 0.2) = 7.2$$

Sure, seen this way, it's quite evident that this technique might not be as efficient as the "classic" technique. For example, perhaps you would have been faster calculating 144 x 5 directly (multiplying by 10 and then halving), and then placing a decimal point "in the right place." However, there are specific occasions where everything changes, and this technique becomes much more useful and effective, making it beneficial to learn to recognize them. Let's say, for example, we want to calculate 8.4 x 2.25.

Here, instead of breaking down 2.25 into (2 + 0.2 + 0.5), let's break it down into (2 + 0.25) and immediately see why. In fact: 0.25 = 1 / 4 and multiplying by 1 / 4 means dividing by 4, so we will have that:

$$8.4 \times 2.25 = 18.9$$
$$(8.4 \times 2) + (8.4 / 4) =$$
$$16.8 + 2.1 = 18.9$$

In practice, I can make this method efficient when I manage to identify (or decompose in such a way as to create) key numbers within the decimal point that transform multiplications into simple divisions; more specifically, these key numbers are:

0.1 - 0.01 - 0.001 - etc. Multiplying by these is equivalent to dividing by 10 - 100 - 1000, etc.

0.2 - 0.02 - 0.002 - etc. Multiplying by these is equivalent to dividing by 5 - 50 - 500, etc.

0.25 - 0.025 - 0.0025 - etc. Multiplying by these is equivalent to dividing by 4 - 40 - 400, etc.

0.33 - 0.033 - 0.0033 - etc. Multiplying by these is equivalent to dividing by 3 - 30 - 300, etc. (note: the "real" number that causes the division by 3 is 0.3 repeating, which means it has infinite 3s. Therefore, the fewer threes you have, the more approximate your result will be).

0.5 - 0.05 - 0.005 - etc. Multiplying by these is equivalent to halving the number - halving and then dividing it by 10 - halving and then dividing it by 100, etc.

0.66 - 0.066 - 0.0066 - etc. Multiplying by these is equivalent to subtracting from the number 1/3 - subtracting 1/3 and then dividing by 10 - subtracting 1/3 and then dividing by 100, etc. (here too the result is approximate, as in 0.33).

0.75 - 0.075 - 0.0075 - etc. Multiplying by these is equivalent to subtracting from the number 1/4 - subtracting 1/4 and then dividing by 10 - subtracting 1/4 and then dividing by 100, etc.

0.8 - 0.08 - 0.008 - etc. Multiplying by these is equivalent to subtracting from the number 1/5 - to subtract 1/5 and then divide by 10 - to subtract 1/5 and then divide by 100, etc.

0.9 - 0.09 - 0.009 - etc. Multiplying by these is equivalent to subtracting from the number 1/10 - subtracting 1/10 and then dividing by 10 - subtracting 1/10 and then dividing by 100, etc.

And so, here are some examples of direct application. I would also like to highlight that "recreating" the presence of one of these "key numbers" can be considered a technique for the more "hardcore" mental calculators, or for those who enjoy a slightly more complex challenge; therefore, if you feel that it doesn't work for you, or that practicing it is not beneficial, feel free to completely discard it from your "arsenal of mathematical tools."

- **25 x 0.8 =** as seen from the key numbers, multiplying by 0.8 is equivalent to subtracting a fifth from the number, so: 25 - 1/5 of 25 = 20

- **30 x 1.8 =** (30 x 1) + (30 x 0.8) = Take 30 and add a (30 from which you subtracted 1/5) = 30 + (30 - 6) = 54
- **30 x 1.809 =** (30 x 1) + (30 x 0.8) + (30 x 0.009) = Take 30, add 30 from which you've subtracted 1/5, and add 30 from which you've subtracted 1/10, then divide the result by 100 = 30 + (30 - 6) + (30 - 3) / 100 = 30 + 24 + 0.27 = 54.27
- **50 x 3.6 =** ? (try calculating it yourself!)
- **55 x 3.2 =** ?
- **2500 x 1.22 =** ?

Percentage decomposition

For all those who understandably might have found the previous method a bit headache-inducing, let's try working for a moment with something slightly different, based on combining decimal calculation and percentage calculation.

Suppose you want to multiply a number by 3.6. In the previous method, you could triple the number and then add 3 times 1/5 of that number. However, there is actually a much simpler way to calculate the same operation by multiplying by 4 and then subtracting 10% from the result.

In fact, if we take the number 4 and reduce it by 10%, we get exactly 3.6. Therefore, given the need to multiply a number by 3.6, we can multiply that same number by 4, then "give back" that 10% which we had previously subtracted from 4, thereby correcting the final result.

To extract a more general rule from what we've just seen, we can say that if we want to multiply by a number with a decimal (like 3.6), and we "notice" that this number is equal to an integer plus or minus a certain percentage (like 4 - 10%), we can multiply by the integer first and then apply the same percentage to the final result. Thus, if, for example, 5.5 equals (5 + 10%), a multiplication by this amount involves multiplying by five and adding 10%; if 7.5

equals (10 - 25%), then you can multiply by 7.5 by multiplying by ten and then removing a quarter from the result. And so on.

Multiplying by 3.6, as some of you may already know, is practically useful for converting speeds expressed in meters per second to speeds expressed in kilometers per hour. For example, suppose you know that a car traveled from Rome to Milan at a speed of 30 meters per second. By multiplying by 3.6 (or multiplying by 4 and then subtracting 10% from the result), you will find that the speed in km/h is 120 - 12 = 108 km/h. However, the reverse, that is dividing by 3.6 to convert from km/h to meters per second, is certainly not as simple. One method, nonetheless, could involve doubling, dividing twice by three, and adding 25% (more cumbersome indeed, but I guess also that… it is what it is).

This principle of breaking down into percentages can give rise to a new "quick multiplication table" with decimal numbers. Again, the recommendation is certainly not to memorize the table (which, clearly, will be anything but exhaustive), but to learn to understand the underlying process and practice recognizing when a decimal number is the result of an integer with a percentage added or subtracted. In these cases, we simply need to multiply by that integer and apply the same percentage.

x1,1 = Add 10%

x1.8 = Double and eliminate 10% (1.8 is 2 minus its 10%)

x2.2 = Double and add 10% (2.2 is 2 plus its 10%)

x2.4 = Double and add 20% (2.4 is 2 plus its 20%)

x2.5 = Double and add 25% (2.5 is 2 plus its 25%)

x2.7 = ? (try to find, as an exercise, a method that uses percentage decomposition to "quickly" solve this type of multiplication)

This method certainly has its drawbacks, as it is perhaps a bit less intuitive compared to the previous one, not applicable to division, and less flexible (with the previous method it was possible to combine the numbers in the table and even carry out complex multiplications like 3 x 0.8225). Nonetheless, it still holds enormous utility, both because of the topic we will address in the next section and because it can still offer significant simplifications whenever these percentages become "recognizable."

Now use these... to perform quick multiplication without decimals!

This chapter on multiplying with decimals can also be extremely useful for calculating regular multiplications quickly... without decimals! For example, suppose you need to calculate 758 x 36. You could use the technique of:

- "Transform" that 36 into a 3.6. This, as we should know well, is equivalent to dividing by 10.
- Calculate x3.6 by multiplying by 4 and then subtracting 10%.
- Multiply the result by 10 at the end to "adjust" for the fact that you had converted 36 into 3.6.

So, you would immediately get 3032 - 303.2 = 2,728.8, which multiplied by 10 equals 27,288.

And you can apply this same method to any number. More generally, in fact, in the case of multiplication between integers, you can:

- Determine if, after adjusting one of the integers, a simpler calculation method can be applied.
- If applicable, move the decimal point by dividing by the appropriate power of 10.
- Execute the simpler calculation technique.
- Multiply the intermediate result by the power of 10 by which you initially divided to obtain your final result.

The traveler's decimals

Imagine now that you are wandering through an English-speaking country, stopping by a stall, and discovering that the painting you like so much is 24 inches tall (or, if you're coming from an English-speaking country already, imagine you have to do the opposite). What do you do to determine if you have enough space to hang it in your living room? The first trick, of course, could be to turn to Google, but what if your smartphone is dead, you don't have a connection, or you simply want to do it quickly by yourself? The most efficient solution, ultimately, given its greater speed and reliability, might be to use the tool contained within your skull. For example:

- **Conversion from inches to centimeters:** one inch is approximately 2.5 centimeters long. Therefore, to obtain the equivalent measurement in centimeters, you can easily take the inches, double them, and add half.

- **Conversion from centimeters to inches:** here you can use the *expressions decomposition:* if 2.5 = 25 / 10 = 10 / 4, then you can easily divide by 2.5 by inverting the 10 / 4, thereby doubling twice and then dividing by 10.

- **Conversion from feet to meters:** one foot is approximately 0.3 meters long. Therefore, here you could simply round to 0.33 and divide by 3.

- **Conversion from meters to feet:** do the opposite and multiply by 3.

- **Conversion from miles to kilometers:** one mile is approximately equal to 1.61 kilometers. If you round it to 1.60, which works just fine for short distances where the approximation isn't too noticeable, you can double and then remove 20%. In fact, 2 - 20% = 1.60.

- **Conversion from kilometers to miles:** as we saw when we talked about the Fibonacci sequence, we can divide by 1.6 using the simple equivalence 1.6 = 16 / 10. Therefore, we could multiply by 10 and then halve four times. Alternatively, if you want something less precise but a bit quicker, you could consider the opposite, that one kilometer is roughly equivalent to 0.62 miles, and multiply by 0.62. To simplify things, you might also approximate to 0.66, which would mean simply removing a third from the kilometer measurement to obtain the miles.

- **Conversion from pounds to kilograms:** one pound is approximately equal to 0.45 kg. So given a weight in pounds, to get the equivalent in kilograms, you can halve it (multiply by 0.50) and then subtract 10% (0.50 - 10% = 0.45).

- **Conversion from kilograms to pounds:** one kilogram is approximately equal to 2.2 pounds. Therefore, to convert, you can simply double the amount and add 10%. Note: here the sequence of operations is the reverse of what was seen in the previous point, but this is solely due to the approximations made. As a general rule, if multiplying by a decimal number involves multiplying by a whole number and then adding or subtracting a percentage, there are no mathematical rules to obtain a "reverse" expression for dividing by the same number. Only the *expressions decomposition* is reversible, so remember that well!

- **Conversion from gallons to liters:** one gallon equals approximately 3.785 liters. Here, you could approximate it to 3.80, multiply by 4, and then eliminate 5%.

- **Conversion from liters to gallons:** one liter equas approximately 0.26 gallons. If in your case it's acceptable to round it to 0.25, simply divide by 4 and voilà, conversion problem solved!

- **Conversion from Celsius to Fahrenheit, and vice versa (exercise).** Here is an exercise for you: to convert a temperature from Celsius to Fahrenheit, you typically multiply the Celsius temperature by 9/5 and add 32. Conversely, to convert from Fahrenheit to Celsius, you subtract 32 from the Fahrenheit temperature and multiply by 5/9. How could this operation be simplified using the techniques we've seen so far?

How to triple your capital

Now, let's apply this technique, combined with those seen in the previous chapter, to a formula that could surely be useful to many of us: if we have an invested capital, in any form, and we want to know precisely how it grows in terms of compound interest (and thereby the reinvestment of profits), the actual formula could be quite complicated to implement and execute. However, a good way to approximate this result could be to apply the so-called "rules of 72 and 115," that is:

- To get a good estimate of how many years it takes to double our capital, *we can divide 72 by our interest rate.*

- To obtain a good approximation of how many years it will take to triple it, *we can divide 115 by the same interest rate.*

Once again: these are rough guidelines, which work best for interest rates that are neither too high nor too low, assume that the interest is always reinvested, and that the interest rate remains constant over time. But let's say the conditions just described are "suitable," you have successfully deposited an amount in an account with a fixed interest rate of 2.75% per year, and your intent is to figure out after how many years it would double, and after how many years it would triple:

72 / 2.75 = Let's try here, simply, to use some *expressions decomposition:* 275 can indeed be decomposed into (5 x 5 x 11) (I'll leave the mental exercise to you to figure out how to calculate this decomposition), and therefore 2.75 is equivalent to (5 x 5 x 11) divided by 100.

Now, we need to divide a number by this "expression," which means "inverting" it, and thus multiplying by 100, then dividing by 5, by 5, and by 11. The steps to calculate this division should therefore be:

72 x 100 = 7200 (trivial)

7200 / 5 = 1440 (dividing by 5: double the number and move the decimal point one place to the left; in this case, just remove a zero)

1440 / 5 = 288 (same as the previous step)

$$288 / 11 \approx 26.18$$

The steps here, furthermore, could have been approximated in various ways:

- Should anyone have had trouble with that division by 11, a trick (never revealed before, ladies and gentlemen!) to approximate it could have been to divide by ten and then remove 10%. Be careful, however: *this is not* a "percentage decomposition" and in no way it offers an accurate result of the division itself, but represents a simple approximation whose "acceptability" of the margin of error must be evaluated based on the use case and some parameters that we will see shortly. In this case, therefore, 28.8 - 10% = 25.92

- Alternatively, we could have rounded 2.75 to 2.8. From which, for example:

$$2.8 = (28 / 10) = (7 \times 2 \times 2) / 10 \text{, and if we follow the inverse expression:}$$
$$72 \times 10 = 720$$
$$720 / 2 = 360$$
$$360 / 2 = 180$$
$$180 / 7 \approx 25.7$$

- Alternatively, we could have rounded that 2.75 up to a 3, resulting in something even further from the actual division result (24), but in exchange for a significant simplification of the calculation process. All the results obtained so far, in any case, suggest a doubling time ranging between 24 and 26 years, which might already start to give us some "very early indications" regarding our possible investment strategy.

The Exercise

Given the nature and divisibility of the number 115, what could be the "most reasonable method to approximate the calculation of the years in which our capital is tripled, given the fixed compound interest of:

3.5% = ?

4.25% = ?

Let's defeat one last "monster"

As we have seen, many of the techniques discussed in this chapter have the main issue of not working for division, a mathematical operation that thus continues to prove itself among the most "unpleasant" of all.

In my opinion, however, we can attempt at least to "tame" this monster through two final techniques that I thought to include in this chapter due to their "inevitable" connection with anything percentage or fractional: repeated subtraction and the (already briefly mentioned earlier, when we discussed division by 11) approximate division.

Repeated subtractions

The first technique involves replacing the division calculation itself with a simpler set of subtractions; and while this may not be definitively classified as a "quick method," it could still be ideal for simplifying the process for those who, for instance, can calculate subtractions very quickly or have significant difficulties with the "classic" column division method.

Some "preventive steps" to be applied before the actual technique are:

- As in classic division, if the dividend (remember: the number before the division sign) is smaller than the divisor (yes, the one after), multiply your dividend by 10 and start placing a "0 point" in the partial result. So, for example, given 7 / 15, you should start writing 70 / 15 = 0.

- If despite the previous operation your dividend is still smaller than the divisor, add another 0 to the partial result after the

decimal point and multiply the dividend by 10 again, repeating this step until the condition is satisfied.
- If there are decimals in the operands, use the invariant property to visually simplify the division and eliminate them. For example, if you have 56 / 0.23, transform it into 5600 / 23; and be careful, because in the following steps, we will refer to this 23 as the "original divisor."
- Now ensure that the dividend and divisor have the same number of digits by multiplying the divisor by a power of 10 that achieves this equality. For example, in the case just mentioned, to make 23 have the same number of digits as 5600, you need to multiply it by 100, so you'll now be working with 5600 and 2300. If, after equalizing the number of digits, the dividend is still smaller than the divisor, remove a zero from the divisor before starting.

And now, the "real technique":

1. Repeatedly subtract your dividend from the new "increased" divisor, do this until the result remains positive and count each time you do it. So, in this case, you have: 5600 - 2300 = 3300, first time. Subtract again, 3300 - 2300 = 1000, second time. You can't subtract anymore, or the result would become negative. In this case, you have subtracted 2 times, and thus 2 is the first digit you need to write in the partial result (to the right of the decimal point, if you have used decimal points in the first steps).

2. Now, start from the result where you left off with the previous subtractions (in this case, 1000), and begin subtracting not from the divisor used so far, but from the divisor divided by 10 (in this case, you will operate with 230 instead of 2300). Repeat the previous steps, counting as long as the number remains positive.

In this example, you would therefore have:

1000 - 230 = 770.
770 - 230 = 540.

540 - 230 = 310
310 - 230 = 80.

Stop here; you have subtracted 4 times and thus can put 4 in the partial result which, for now, is 24.

3. If the number you've stopped at (in this case, 80) is still greater than the original divisor (in this case, 23), repeat the previous step, starting with the number you stopped at after the previous subtractions, and begin subtracting from it the divisor, again divided by 10 (so instead of 230, start subtracting 23). So, in this case, you have:

80 - 23 = 57
57 - 23 = 34
34 - 23 = 11

And you stop at 11 since subtracting further would make the number negative. You have subtracted 3 times, so you can place 3 in the partial result, which is 243. Attention: now you have a number, 11, which is also less than the original divisor of the entire operation. Therefore, the partial result, 243, is also the final result. The number you stopped at, however, is indeed the final remainder of the division.

4. If you then want to proceed to calculate the decimals, you just have to keep dividing the divisor by 10 and continue with the subtractions as in the previous steps. For example, 23 / 10 = 2.3 and therefore:

11 - 2.3 = 8.7
8.7 - 2.3 = 6.4
6.4 - 2.3 = 4.1
4.1 - 2.3 = 1.8

You subtracted four times, and indeed 4 is the first decimal digit.

And you can continue in this manner, perhaps starting by subtracting 0.23 from 1.8 until you deem it appropriate.

Sometimes, numbers with an infinite amount of decimal places can emerge, so it's necessary to stop simply when you have enough relevant digits.

Approximate division

A few pages ago, we "suddenly" introduced the trick of approximating a division by 11 by dividing by 10 and then removing 10%. This principle can actually be expanded into a method called "approximate division" that works as follows:

Think of the number you want to divide by, let's call it B (for example, 123). Find an easier number for the division, let's say C, that is the closest to B (perhaps 100, as it is "close enough" to 123).

Now, if you divide any number by C instead of B, this will obviously give you an approximate result. This might even be acceptable in some practical cases, but the truth is, it could be much more useful to make "a small correction" first, as in the case of 11.

This correction will obviously depend on the distance or difference between B and C. In the specific case of dividing by 100 instead of 123, we've substituted a larger number with a smaller one; therefore, the result will be larger (obviously: a cake divided into fewer slices will have larger slices) and to correct it at least partially, we will need to "subtract" something. For example, if we had decided instead to divide by 200 instead of 198, we would have needed to add something at the end, a new value "Y." In short: an increased approximation in the divisor will require a negative correction Y, and vice versa.

In this case, therefore, just like in the case of 11 from a few chapters ago, a "sufficiently" valid rule could be to make this "correction" Y the percentage difference between B and C. In the case of 11, it was 10%, here this value corresponds to about 25%.

From which, providing a practical example:

$$7600 / 123 \approx \text{ (where "}\approx\text{" is the symbol for "approximately equal to")}$$
$$(7600 / 100) - 25\% = 57$$
$$76 - 25\% = 76 - 19 = 57$$

Note that the actual result would have been approximately 61.79, which, unlike in the previous case, results in a much larger discrepancy from our value. This highlights the (obvious?) rule that this approximation will be greater where the difference between B and C is wider: the further C is from B, the more our approximation will diverge from the actual result. In our example, the difference between 123 and 100 is "relatively wide," which caused a significant discrepancy in the approximated result, potentially making the entire process not particularly useful.

Beyond assessing the cost/benefit ratio case by case for this procedure, I feel compelled, before concluding, to take a step back and view the practices discussed in this chapter from a broader perspective. All the discussions so far on approximating calculations, aside from making the mathematical method more creative, provide extraordinarily valid working principles to "train" us to think in terms of cost/benefit. I believe this can help us develop an ideal mindset to assess when, not only in mathematics but in any field, going beyond a certain degree of precision might incur costs that are no longer worthwhile or sustainable. Alternatively, if we wish to "invert" the concept, when small approximations in our goals, or expectations, can lead to substantial savings in terms of time, money, and resources. I am not sure if there already exists some physical or mathematical law that describes this phenomenon, but everyone has likely noticed that, in any project or endeavor, a genuinely *"reasonable threshold of imperfection"* can be easily identified; a demarcation line where further effort in accuracy, completeness, or control is not justified by the associated costs or sacrifices. Hence, in my opinion, trying to pinpoint this threshold whenever we aim to define our objectives or our paths remains one of the most precious tools of "intelligent productivity" we can ever have; a true "master mindset," steering us away from the obsession with "results at all costs," towards a much more sustainable pursuit of "achieving the best possible in exchange for the most reasonable expenditure."

The Historical Pill

Carl Friedrich Gauss is considered one of the greatest mathematicians of all time, the "father" of fundamental contributions such as the fundamental theorem of algebra and the normal distribution in statistics (which we will see, in part, at the end of this book). Gauss was born in 1777 in a family of humble origins in Germany and showed a talent for mathematics from a very young age. It is said, for example, that at just three years old, he corrected an error in his father's payroll count. At the age of seven, while in school, his teacher gave the class an activity to sum the numbers from 1 to 100, thinking it would keep the students occupied for a while. However, Gauss solved the problem almost immediately, realizing that he could group the numbers into pairs that summed to 101 (1+100, 2+99, 3+98, etc.), for a total of 50 pairs, which equals 5050.

The Aphorism

"The beauty of mathematics lies not only in its answers but also in its questions. It teaches us to think, to reason, to ask the right questions. And sometimes, the questions are more valuable than the answers themselves."
(Grigori Perelman)

Mathematical Oddities

Another paradox generated by the concept of infinity is linked to set theory. The initial idea is to consider the set of all sets. In theory, this should be possible, but in reality, this assumption creates a paradox, known as Russell's paradox.

If the set of all sets actually exists, then it should include itself as a member (because it includes all sets). But then it should also include all sets that do not include themselves, which creates a "no exit" cycle of unsolvable contradictions.

XVII - A Game of Chances

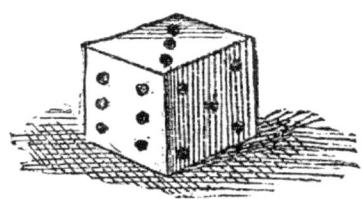

Arthur Benjamin, a prominent mathematician, stated during a TED conference that instead of focusing on mathematical analysis, every school curriculum should, in his opinion, focus on the study of probability theory, as it is a field that is simultaneously more fascinating, useful, and enjoyable. According to him, a greater widespread knowledge of statistics and probability would almost certainly have prevented the financial crash of 2008 and helped everyone make far wiser decisions than those responsible for such an economic earthquake (and, I might add, decisions that we tend to unfortunately repeat cyclically as humans).

But beyond Benjamin's personal opinion, which we can debate as much as you like, it is, in my view, undeniable that gaining a better understanding of statistics and probability theory can provide us with extremely powerful tools. These tools help in assembling past data, gaining a greater understanding of the present, grasping the true nature behind the evolution of ideas and phenomena, and even improving one's hypothetical forecasting abilities for the future. Probability, in fact, is one of those mathematical phenomena that is often highly counterintuitive, and understanding its dynamics therefore adds to the arsenal of fundamental tools with which to gain a true competitive advantage.

Let's consider, for example, how people tend to be extremely frightened by very unlikely yet narratively captivating ideas, such as the possibility of a plane crash or being eaten by a shark. Yet, the odds of dying in a plane crash in 2020 were the same as flipping a coin 21 times and getting heads every time (try it, and you'll understand why they say that airplanes are the safest means of transport in the world). Meanwhile, it's clear that sharks kill fewer people compared to horses, mosquitoes, or yes, even the accidental fall of coconuts.

Let's be clear, probabilistic data alone are not, and never will be, enough to "completely eradicate" certain fears or irrational concepts that remain part of us, as they are more compatible with the very configuration of our brain. Our "biological command center" is generally much more inclined to perpetuate a mindset like: "If something is possible, however improbable, I could still be the unlucky (or lucky, in the case of a lottery) one to encounter this extreme improbability." And this reflection, to be clear, is theoretically true, but the factor to focus on here is: why do we think this way only about fascinating, noisy factors that hold some emotional relevance for us, while we typically tend to ignore more "silent" but also more impactful factors? (consider someone who avoids flying believing it extends their life, but then calmly smokes three or four packs of cigarettes a day).

And above all, is it really worth "paying" the required cost, even if it's emotional, to consider these extreme improbabilities? Given that our attention is a limited resource, isn't it more beneficial to redirect our focus toward "more probabilistically relevant" factors, even if they're less "fascinating," for our finances, our goals, our survival?

The journey towards a mindful navigation of uncertainty can, in essence, be quite complex and may involve a detailed analysis of "cognitive biases" and psycho-neurological foundations, which, for the sake of simplicity, we won't cover in this book. Nonetheless, I strongly believe in the importance of informing and educating in a "mathematical" way, at least as a means to equip ourselves with the tools to embark on such a journey. This approach allows us to reorient our most impulsive and instinctive responses towards more rational and informed decisions. Whether or not this leads to a path of self-awareness will largely depend on the common sense of each of us.

And so, after having covered all the philosophical preliminaries, let's dive straight into the essence of the subject. We'll begin with what, in my opinion as well, remains one of the most fascinating and useful applications of mathematics. We'll start with what is perhaps one of the most "pragmatically relevant" indices in the field of probability: the expected value index, or expected value.

Expected value, and two ways to predict the future

The expected value index can be understood as the average gain (or loss) that we obtain if we continue to perform certain actions in contexts of uncertainty. In other words, it could be the numerical answer to questions such as: "On average, how much money do I gain or lose each time by continuing to bet three dollars on red in the roulette game?" or "What can I expect each month if I continue with this investment strategy?"

The formula to calculate this value is actually quite simple, and it can be formalized as:

$$(\text{Probability of gain} \times \text{Gain}) - (\text{Probability of loss} \times \text{Loss})$$

Once we've calculated this index, if the value is positive, continuing to "do that thing" will yield a consistent gain in the long run. If it's negative, we will incur a loss.

This is a value whose calculation can be extraordinarily useful, whether you want to determine if a decision influenced by uncertainty is actually advantageous in the long term, or whether you want to understand, among multiple choices, which one might bring us greater gain, or perhaps the least loss (since, in that case, we "just" need to choose the one with the highest expected value).

However, I know what many of you here might be thinking: if every situation in our lives could be so easily modeled, we would have already solved a myriad of problems long ago. The first complexity, in fact, that arises from the suggestion to apply this formula is: how do you calculate the probability of an event occurring? Well, mathematics offers us two precise answers, each with its own clear advantages and disadvantages:

- **A method consists in a statistical procedure**, which involves looking at the outcomes of similar events that occurred in the past, and then asserting that the probability corresponds to the number of times the event happened divided by the number of times it "could have happened." For example, from a statistical point of view, although highly simplified, one might claim that the probability of an airplane having an accident is equal to "Airplane accidents that occurred" divided by "Total flights."
This method obviously presents a significant initial disadvantage, a real epistemic limit (that is, a barrier in the ability to understand reality) due to the fact that even after calculating such a datum, it is not always (or almost never?) possible to predict a future event's trajectory by gathering past data; in fact, often this practice can be the source of disastrous decisions. Thus, the statistical estimate of probability should be understood primarily as something descriptive, rather than predictive of a phenomenon's uncertainty; that is, a datum that indicates how "something has been" so far, more than how "it will be from now on." Beyond this limitation in our ability to predict the future, however, it could be said that in most cases, for our predictions, it "may be convenient" or "we will inevitably rely" on what existing data up to that point seem to suggest to us. Consider, for example, the possibility of a total and sudden upheaval in the economic or even biological laws of the world around us: nothing guarantees that such upheaval is not always around the corner, but it is still true that, in the majority of cases, it will be more "economical" for us to proceed with a sort of "illusion of stability," the idea that things will remain "more or less" similar to the past. This "illusion of stability," after all, despite "distorting" reality somewhat, is what allows us to leave home without constantly thinking about being struck by meteorites, marauders, pirates, or a sudden subversion of physical laws that completely reverses gravity; this

simplification is, simply put, an essential part of what allows us to function.

If one were to actually think about improving the descriptive value (in practice) or predictive value (theoretically) of their statistical data, this can always be done by collecting "higher quality data." For instance, in the previously mentioned case of airplane accidents: does it really make sense to consider the flight data from 1920s to understand my probability of flying safely? And those of a plane from another company? And those of a completely different model? These questions obviously open up an infinite number of dilemmas, typical of "data science" and sector specialists, and therefore are not worth addressing in this text. For now, then, let's try simply to keep in mind the basic formula:

Number of times it happened / Total number of times it could have happened

and always consider it along with a "necessary evaluation" of all the complexities involved in data collection that it may imply.

- **Another way to determine the probability of an event is through the theoretical/mathematical method**, which relies on the analysis of the system or process involved and the consequent development of precise equations. Returning to the example of an airplane crash, such a method could lead us to extract the probability of an accident for a specific aircraft by evaluating and mathematically combining the probabilities of all the events capable of causing a problem: the "failure" probability of each set of vital components, combined with the probability that the crew might not be able to handle the emergency, and so on.

As can be imagined from the example just given, this probability calculation method applied to complex phenomena can result in incredibly intricate systems of equations. In this chapter, therefore, while focusing more specifically on the theoretical/mathematical method, we will do so within the realm of more immediately understandable phenomena.

Dice, coins and random stuff

A foundational formula worth addressing is the one stating: "If an event can occur in x possible ways and these are equally likely, the probability of one of these events happening is 1 / x."

So if we flip a coin, and it can land on heads or tails (an event that can therefore happen in 2 possible ways), the probability of one of them occurring, for example, landing on heads, is 1/2, that is, one in two, or 0.5.

The same applies if we roll a six-sided die: it can land on a number from 1 to 6, resulting in an event with 6 possible outcomes, and the probability of any one of them occurring, say rolling a 4, is 1/6. This is probably very simple, if not already known, to most of you; let's try to take a "small step further."

If we are faced with the same event that can occur in x possible ways, all with the same probability, and we want to know the probability that it happens not just in one way, but in any one among two, three, etc. of those ways, the probability will be exactly 2 / x, 3 / x, etc.

So, if for example we have a die, and we want to know "the probability of rolling a 2, 3, or 4" (the probability that an event with 6 possible outcomes results in one of 3 desired ways), the probability will be 3 / 6 = 1 / 2. It would have been 2 / 6 = 1 / 3, if only the 3 and the 4, or any other pair of possible results, were of interest to us. The rule, in short, is always the same: we need to divide the number of "events we are interested in" by the total number of events.

This explanation, in theory, also suggests what to do when there are multiple possible events, which are not equally probable. One should always try to think as if facing a pie, whose "whole" represents 100% probability (or probability 1, or certainty) — the probability that at least one of the possible events will occur. Logically speaking, given all possible scenarios, the phrase "at least one of them will happen" offers a certainty. Then, this pie should be "divided" among these possible scenarios according to how probable they are in comparison to others.

In the case, for example, of a loaded die with an internal weight that makes a 6 come up 40% of the time, we would have (in a scenario that is clearly simplified, since a weight that makes a 6 appear so often would also greatly reduce the probability of a 1 coming up):

- Probability of rolling a 6: as mentioned, 40%, or 0.4
- Probability that any other number comes up: 60% (the remainder of the "pie already taken" by 6), equally divided among the remaining 5 numbers, which are equally probable = 12%, or 0.12

Here, it may be necessary to add a small note for those who may not be entirely clear on the notation we used to express probabilities: so far, we have discussed probabilities expressed not only in percentages but also in numerical terms. This is necessary because, as we will see, it tends to simplify certain arithmetic operations. However, just to clarify: a probability can always range between 0 (an impossible event, or 0%) and 1 (a certain event, 100%). The conversion between the two notations is quite simple; it involves dividing (if you want to convert from percentages to numbers) or multiplying by 100 (for the reverse). So, for example, a 25% probability can be expressed as 0.25, a probability of 0.999 indicates a 99.9% probability, and so forth.

Negations, coincidences, and mutual exclusions

Three more mathematical formulas that could prove extremely useful for solidifying our understanding of probability theory are:

- **Formula for non-occurrence (NOT):** The probability that something does not happen is simply 1 minus the probability that it does happen. For example, if the probability of rolling a 6 on a die is $1/6 \approx 0.17$, then the probability that a 6 does not appear is $1 - 0.17 = 0.83$.
- **Formula for coinciding events (AND).** The probability of independent events happening together (assuming for simplicity that the possible occurrences cannot influence each other) is

obtained simply by multiplying their probabilities of occurring. For example, if the probability of flipping a coin and getting heads is 1/2, the probability of flipping two coins and getting heads both times (or any other combination) is:
1/2 x 1/2 =
1/4 = 0.25, or 25%.

- **Formula for independent events (OR).** The probability that, given multiple independent events, at least one of them happens in a certain way is given by 1 minus the probability that NONE of them happen coincidentally. At first, this might seem like a bit of a tongue twister as phrased, but when you think about it, the origin of this formula is quite intuitive: "at least one of these things will happen" is logically identical to "it is NOT true (hence 1 -) that none of them will happen (thus the product of the NOT probabilities of the individual events)."

This case might seem a bit more complex, but again, I believe an example can easily clear up any doubts. Suppose we roll a die and flip a coin and want to know the probability of getting heads on the coin flip or rolling a 6 on the die. Therefore we must calculate:

- **Probability of not getting heads:** 1 - 1/2 = still 1/2.
- **Probability that the die does not show 6:** as discussed, 5/6.
- **Probability that both things do not happen:** thus the product of the two = 1/2 x 5/6 ≈ 0.42.
- **Negation of the previous:** 1 - 0.42 = 0.58, which is precisely the probability that we get heads on the coin flip or roll a 6 on the die.

Casino and multiple choices

Having reviewed everything needed to understand at least some essential basics of probability calculus, let's immediately look at a practical example of the usefulness of the expected value formula by analyzing the case of American roulette.

As many might know, American roulette consists of a spinning wheel divided into 38 slots, 18 of which are red, 18 black, one zero, and a double zero. A ball spins on the wheel, and each turn, you can bet money on the number (or its color, parity, etc.) where the ball will land.

Let's say, then, that the action for which we want to calculate the expected value is repeatedly betting 1 dollar on red.

Let's look at the formula for the expected value, and remember that it is (Probability of gain x Gain) - (Probability of loss x Loss); so, let's try to understand what the various values to insert into the formula are:

- **Gain** = 1 dollar, because that's what we will win if red comes up.
- **Loss** = 1 dollar, because that's what we will lose if the ball lands on black, zero, or double zero.
- **Probability of gain** = Since there are 18 red slots, by always betting on red we should win 18 times out of 38. We use the second method of calculating probability. We have an event with 38 possible outcomes and we want to know the probability that one among 18 of these possible outcomes occurs, so we simply calculate "Events of interest / Total events" = 18 / 38 = 0.47 (or 47%).
- **Probability of loss** = With 18 black squares and 2 green squares, if we always bet on red, we will lose 20 times out of 38, resulting in 20 / 38 = 0.53 (or 53%). And already here, the fact that the probability of loss is slightly higher should start to seem suspicious to us.

Applying these values to the expected value formula, we have:

$$(1 \times 0.47) - (1 \times 0.53) = -0.06$$

What does this negative number tell us, then? That for every dollar bet on red, we will have an average loss of six cents. And you can very easily verify that the value won't change if we replace red with black. Therefore, inevitably, if we continue playing roulette, regardless of our strategy our wallet is destined to shrink more and more in the long term. After all, as many of you already know, it is precisely because of this negative expected value that casinos enrich themselves at the expense of those who keep playing. Except for some games where skill can make a difference, like Poker or BlackJack, the only way to win at the casino is to play as few times as possible with minimal amounts, to "reckon" with the result of these few plays, whatever it may be, and... run away!

But let's move beyond that and see how to tackle this tool in fields other than gambling. For example, to improve our performance in multiple-choice tests.

Let's assume we have a test with multiple questions, each of which can be answered by choosing from five possible responses: A, B, C, D, E. Suppose that each correct answer earns us 3 points and there are no penalties for incorrect answers. Additionally, let's assume we don't know the correct answer to some of these questions, and therefore we want to "guess" for those. Thus, let's refer to our expected value formula and see exactly how we should proceed:

- **Gain =** 3 points
- **Loss =** 0, because there are no penalties for an incorrect answer
- **Probability of gain =** Since there are five answers, the probability of answering correctly by guessing is simply one in five, or 1/5 (probability calculated using the second method, equivalent to 20%).
- **Probability of loss =** the likelihood of answering incorrectly by guessing is 4 out of 5, which is 80%, but we'll see that since the loss equals 0, its probability doesn't matter; in fact:

$$(1 / 5 \times 3) - (4 / 5 \times 0) = (1 / 5 \times 3) - 0 = 3 / 5$$

The expected value here is positive, meaning we will have an average gain of a few points (3/5 of a point, which is still better than nothing) for each randomly guessed answer. Therefore, if there is no penalty for wrong answers, it is always advantageous to guess randomly if we don't know the response.

But now, let's introduce a penalty of 2 points for incorrect answers in the same test. Therefore, the formula becomes:

$$(1/5 \times 3) - (4/5 \times 2) = 3/5 - 8/5 = -5/5 = -1$$

That is, with such a penalty, the overall expected value becomes negative, and the risk we run is to lose, on average, one point for every answer guessed at random. Now it suddenly makes more sense to leave the questions we don't know the answers to blank.

However, it would be different if we managed to eliminate three possible answers by exclusion and take a guess between the remaining two. In fact, in that case, the probability of getting it right would increase from 1/5 to 1/2, and the probability of getting it wrong would decrease from 4/5 to 1/2. Thus, we would have:

$$(1/2 \times 3) - (1/2 \times 2) = 3/4 - 2/4 = 1/4$$

which again is a positive expected value. The presence of a penalty alone, therefore, is not enough to say that it's no longer advantageous to guess randomly. So how can we more rigorously determine when it is actually beneficial to guess randomly in multiple-choice tests with penalties?

Developing the expected value formula through steps that, for simplicity, we won't detail, the answer is "Take a guess if":

Points per correct answer > Penalty points x (Total responses - 1)

That is, we multiply the penalty points of a wrong answer by (Total possible answer choices - 1), and we check if the resulting amount is greater or lesser than the points awarded for a correct answer. If it is, it's better to guess randomly when you don't know an answer. Otherwise, it's better not to answer.

To further elaborate on the topic, you could proceed, in the event of a penalty, with the following strategy:

- When you can eliminate definitely wrong answers from a question, do it regardless, because it will increase your expected value and, consequently, your average gain in the case of a random answer.

- If the points awarded for a correct answer are actually greater than the penalty points multiplied by (1 - total responses), you can safely make a random guess and not worry about it anymore.

- If not, reduce the "total answers" by one and see if this time the points given by the correct answer are greater than the other amount. If so, it might be worth guessing; if not, recalculate by reducing the "total answers" by two, and repeat the same procedure until (if possible) trying becomes beneficial again.

But before concluding this specific topic, let's extend the formula that answers the question "When to randomly guess an answer?" to all gambling situations, transforming it into a general "When is it advantageous to take a gamble?." Let's develop the expected value formula step by step (not detailed here). We'll see that the condition for deciding whether it's beneficial to take a repeated gamble depends on:

Probability of Gain x Gain > Probability of Loss x Loss

In its simplicity, this formula represents an invaluable asset because, when used correctly, it can help you in ways you can't even imagine. For example, when they say "you have nothing to lose" by asking that girl or boy out, it means that it doesn't matter whether you are accepted or rejected, because if the loss is negligible or irrelevant compared to the potential gain, a bold and enterprising attitude will always have a positive expected value in the long run, therefore bringing you only gains in terms of well-being, happiness, and personal success. This perhaps also reminds us a bit, from a mathematical perspective, of the value of resilience: if something harms us largely to the extent we allow it to do so, working towards better emotional management in the face of failures, as well as developing an "evolutionary vision" that transforms obstacles into resources, will enable us to "significantly reduce" the value on the right side of the inequality just mentioned. And this gives an even more complete and mathematical sense to the idea that "with the right attitude, we can gain from some risks regardless of the actual factor of uncertainty involved."

When to insist on "Spinning the drum"?

So far, for the sake of simplicity, we have only analyzed situations where previous events do not influence subsequent ones. However, let's consider a case, hopefully very unlikely, where this influence is present, and try to understand how we can use it to "save our lives."

In the "game" of Russian roulette, for example, which involves placing a single bullet in a six-chamber revolver, spinning the cylinder, and pulling the trigger while aiming at your own temple "hoping" to survive (a scenario you've hopefully only experienced in movies), the person who plays last has a much higher chance of dying, but only if the cylinder isn't spun again after each attempt. For instance, if you are the first to shoot, you have "only" a 1/6 chance of hitting the bullet; however, if by pure chance you are the sixth player, and no one has hit the bullet before you, then the probability of getting it at the end is, literally, 1 in 1. This is a clear case of "conditional" probability, where previous events influence those that follow, similar to a die that "loses a face" with each roll, or in the example of drawing numbers from an urn: "sooner or later," if you keep drawing, your number must inevitably come up.

However, there is a way to prevent the odds of dying from increasing with each shot, and that is to "spin the chamber"; each time you spin the chamber, it's as if you are "resetting the game."

It no longer matters where the bullet was before: when the chamber is spun, the bullet will stop "inching closer," and each chamber will once again have a 1 in 6 chance of containing the "fatal bullet."

This, to be clear, doesn't mean in any way that you should risk your neck in such a foolish game; however, the story might lead us to reflect on and help us better negotiate all the instances where the occurrence of something improbable yet significant is actually just around the corner and draws nearer as time passes. And, to be clear, this includes positive improbabilities, such as winning a lottery or getting a "yes" to an important job contract.

This "slow advancement" of improbability, the singularity in every situation, actually occurs even in the absence of conditional probabilities and is due to a phenomenon that anyone with some knowledge of probability knows well as the "law of large numbers": a phenomenon that states that "every possibility will eventually get its turn if stretched over a long enough period." It is the reason why if low-ranked team A has a 10% chance of beating champion team B, it is almost certain that "sooner or later," over many repeated matches, the outcome will defy all our expectations.

And so this principle, in synergy with the previous one, can lead us to a primary, important strategic-tactical tool: if our desire is to achieve a "positive improbability," the most natural response is to repeat our attempts as much as possible, because "sooner or later," success will arrive, even if our chances are very low, provided we play long enough. This, of course, imposes several "buts": the probability might not just be very low but actually zero, in which case no number of attempts will genuinely lead us to success. Or, it might be so low that it requires a number of attempts we will never have the time to make. Above all, it requires us to pay careful attention to the expected value, and thus, to the possibility that the average cost of a single attempt might still make the endeavor fruitless (as in the case of roulette, where, yes, we will certainly "win eventually," but we will always end up with less money than we started with).

However, what we have seen so far also suggests that we can aim to achieve what we need in reasonable time frames by increasing our "conditional probabilities" and consolidating the "historical factors" that bring us closer to our goal. This means trying to "never spin the wheel," ensuring that each previous attempt actually influences the next one positively. This could be achieved, for instance, by analyzing and understanding the factors that negatively affected each attempt, and then applying the "lessons learned" in future situations.

At the same time, if the goal is to "avoid a negative improbability," akin to Russian roulette, our optimal strategy might be to "spin the drum as much as possible," meaning to mitigate all historical factors from our context that could make this improbability "ever closer." Consider, for example, deciding to embrace a truly healthy lifestyle in all aspects: the very "reset" of our vital values that ensues could increasingly distance us from the risk of diseases which, in the past, we have heightened through much more reckless choices. The same applies if we need to propose our idea to someone willing to invest money: if our goal is to "spin the drum," we could decide, after a relentless series of failures, to muster the courage to explore new, brave, unexplored paths. Notice the symmetry in both cases: the drum symbolizes the memory of physical things, information, and probabilities that slowly accumulate, bringing us ever closer to the "singularity": if the singularity is desirable, we should favor this accumulation; if not, we should try to reshuffle the cards and promote the literal "zeroing out" of this memory as much as possible.

This reasoning, ultimately, should lead us to reflect on how, if it is true that we find ourselves in a universe where disorder is the norm—and thus the "Russian roulette of negative improbabilities" is always lurking—the only way we can protect ourselves is by being proactive, by not allowing ourselves to be swept away by this constant progression towards disorder. Therefore, we must generate a continuous balance by experimenting, retrying, and continuously "playing" with reality. This is the only true way for the most beautiful and rare things in life to appear, manifest, and begin to flow harmoniously into our lives.

The exercises

- Could you tell me the probability that, in any given school class, two students were born on the same day of the year (assuming, for simplicity, that they were all born in the same year)? Could you guess intuitively? And could you, instead, develop a series of mathematical steps to justify an answer, even if approximate?

- In 2004, it was calculated that the asteroid Apophis 99942 had a 2.7% chance of colliding with Earth in 2029. Considering the catastrophic event that would have ensued, this was a rather high probability (about 1 in 30), similar to rolling a die twice and getting a "1" both times. Fortunately, subsequent calculations lowered the probability from 1 in 30 to 1 in 230,000, which is the same probability as rolling a die seven times and always getting a "1." However, if we've paid attention to what was discussed in the previous chapter, we should also have understood that rolling a die seven times and always getting a "1" actually has the exact same probability as rolling it seven times and getting any other combination of numbers. So yes, even now, if you take any die and roll it seven times, you'll get a combination of results that, if calculated beforehand, would have had the same probability as the asteroid Apophis wiping us all out in 2029. Therefore, if it's something so "common" and reproducible, why hasn't anyone thought to associate 2029 with a "certain apocalypse"? (and no, "it's all a conspiracy to avoid spreading panic" is not an option). Where is the "flaw" in the reasoning, presented in this manner?

The Historical Pill

Ada Lovelace (1815-1852), daughter of the poet Lord Byron, distinguished herself with her passion for mathematics and science, earning a reputation as the world's first programmer. She collaborated with Charles Babbage on his "Analytical Engine" project, a theoretical mechanical calculator that was never built, and wrote a series of detailed notes on its operation, including what some recognize as the "first algorithm." Lovelace had a forward-thinking vision that mechanical computers would become "more than just mathematical calculation tools," although she dismissed any likelihood of unforeseeable properties emerging from them, a notion that might have been disproven with the arrival of artificial intelligence, for instance.

The Aphorism

"I like to view mathematics more as an art because the activity of mathematicians, who constantly create, is guided but not controlled by the external world of the senses; thus, it actually resembles the activity of an artist, of a painter."
(Bocher)

Mathematical Oddities

The "Monty Hall Paradox" is one of the most debated concepts in the field of probability mathematics due to its counterintuitive nature. For those unfamiliar with it, imagine being on a game show with three closed doors: behind one of them is a car, while behind the other two are goats. You start by choosing a door, and then the host, who knows what is behind each door, opens one of the other two doors, revealing a goat. Now you are given the opportunity to change your initial choice. Should you do it? Mathematics says yes, because the probability that the car is behind the door you initially chose is 1/3, while the probability that it is behind the other remaining closed door is 2/3. Indeed, the mere fact that the host, knowing the contents of the doors, deliberately opens one with a goat, removes one of the "undesirable" possibilities from the game, thus changing the probability distribution. The same scenario might appear more intuitive if there are 1000 doors: the first choice has a 1/1000 probability of winning. Then, if the host opens 998 doors, all with goats, the probability that the car is behind the last remaining door is 999/1000. Think of it as the gradual opening of doors being similar to a "bullet getting closer" in the case of Russian roulette, and you might be able to mentally "solve" the paradox.

XVIII - Estimation and Casting Out Nines

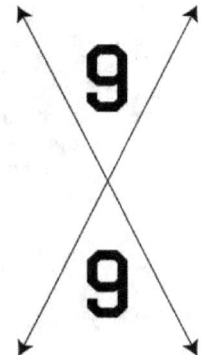

Estimation and "casting out nines" are two extremely valuable tools for error verification (and beyond) in both mental and written calculations. Clearly, both require some time, albeit brief, to be applied. This means that integrating them into our calculation procedures will always incur an *unavoidable time cost* for increased accuracy. I also find that, after a period of relative insignificance, at least in daily life, tools like the "casting out nines" are ready to experience a new, albeit perhaps brief, "phase of glory." In my opinion, this is due to the success of large language models like ChatGPT and their occasional imprecision in performing certain mathematical operations (given that they mostly "guess" the next word in a sentence rather than going through genuine mathematical or logical reasoning). In this "unprecedented" context, indeed, forcing (through proper prompting) these models to employ such cross-verification tools for calculations might reduce their margin of error, which can make a difference in cases of repeated or extremely complex operations. Or at least, this is a potential solution to what, as I write, is an actual technical problem but obviously, as you're reading, the problem in question may have already been resolved.

But let's get back to our two mathematical tools. More specifically, we have that:

- **Estimation** is a tool that can be applied before calculating any operation. It involves creating a faster approximation of the calculation and can allow us to quickly understand what the result will "more or less" be. Therefore, if the final result is very different from the initial estimate, it's quite likely that we've made some mistakes during the process. However, it's clear that if there are only relatively small errors, the estimate won't provide us with useful information. Additionally, this technique has the incredibly valuable advantage of serving as a substitute method for the original calculation when exceptionally high degrees of precision are not required.

- **The "casting out nines" method** is a tool that can be applied after performing any calculation to relatively reliably indicate whether an error was made during the process. Interestingly, evidence of this technique dates back to the 3rd century AD, where it was known as *abjectio novenaria*. However, considering the Romans used pebbles, abacuses, and a numeral system quite different from ours, it is highly unlikely that they understood why it worked. The first documented explanation of the method's functionality comes much later, specifically in 1202, with Fibonacci's *Liber Abbaci*, the same work that helped popularize Arabic numerals in Europe.

Let's begin by discussing estimation, first distinguishing between the three types of estimates that can be made:

Estimate: rounding with zeros

Rounding with zeros is a technique that can be used when working with whole numbers, typically "large" ones, and involves replacing some (or all) of the rightmost digits of the number in question with zeros. However, this alone is not enough. To make the resulting estimate a bit more consistent and precise, it would be better to also increase by 1 the rightmost of the non-replaced digits, if the leftmost of the replaced digits is greater than 5. For example, before multiplying 7,845 x 9,871, we could:

- Replace the last three digits of the first number with zeros to get 7,000. Then, since the leftmost of the replaced digits was an 8, obviously greater than 5, round it up to 8,000.

- Replace the last three digits of the second number with zeros, resulting in 9,000. Then, since the leftmost among the replaced digits was still an 8, increase it to 10,000.

- Calculate then 8,000 x 10,000 = **80,000,000**

The actual result of 7,845 x 9,871 would have been **77,437,995**, and as we can see, 80,000,000 is already a decent approximation. In fact, if at the end of the "actual" multiplication, we had obtained a result in the range of 20 or 200 million, we could have immediately understood that something went wrong during the procedure. Clearly, for more refined estimates, we could have kept not only the last digit but also the second to last, estimating the operation not with 8,000 x 10,000 but with 7,800 x 9,900 (still quite quick to calculate using any of the techniques we have seen on previous pages). The result of this latter operation is indeed 77,220,000, which is even closer to the actual result.

In summary, it should be fairly simple: the fewer digits replaced with zeros, the more refined the estimate will be, and the longer and more complex the calculation will become; yet at the same time, the result will be closer to the actual value. The ability to "seek this balance" by adopting more or less precision is in itself an extraordinarily powerful mathematical tool, since, as mentioned earlier, the "acceptance" of a certain degree of error, where the situation allows, will allow us to reduce the complexity of our calculations by several orders of magnitude.

As some might have already noticed, this is somewhat the "integer equivalent" of the *rounding for fractional numbers* explored a few pages ago; clearly, the assumption that differentiates these two cases is that in the previous instance, given the presence of "generally small" numbers, certain degrees of approximation become much more "acceptable," to the point of making approximations themselves (and thus "substitution with an estimate") a necessary part of the calculation. Consider, for example, the case we just looked at where we approximated 7,845 x 9,871: by "cutting" three digits, we ended up with a difference of almost three million between the estimate and the actual result, whereas in that case, we were discussing errors of just a few cents, thousandths of value, or less. It's clear that we might encounter practical situations where these tenths and thousandths can indeed make a difference but, once again, everything lies in cherishing that "precious rule" of always assessing the right balance between choosing a degree of precision and the actual cost of that choice.

Estimation: replacement with fractions

This technique offers improved estimates over earlier methods by simplifying calculations using "rounder" numbers. Rather than just replacing digits with zeros, it allows us to substitute one of the numbers with a fraction or multiple of 10 that is closer to the original value. This approach keeps calculations easy while giving us the flexibility to round and estimate with numbers that more accurately reflect the originals, thereby reducing errors.

This generally presents the same difficulties encountered in the *expressions decomposition* seen in Chapter IX. It is clearly not an immediate operation to perform and it is recommended only for the more experienced. However, once you have trained your eye a bit to recognize these patterns, it is likely that you will also gain a lot in terms of quality, speed, and accuracy of your calculations. The table that follows can certainly help us "train the eye" for this kind of operation:

- **142 = approximately 1000 / 7.** Therefore, if you find yourself having to multiply 148 by 344, you could round 148 to 142 and then use that 142 in a multiplication by 1000, followed by a division by 7.

- **333 is approximately 1000 / 3.** So, in the previously mentioned example of 148 x 344, you could round 344 to 333 and the latter to 1000 / 3. Thus, you would transform 148 x 344 into (1000 / 3) x (1000 / 7) = approximately 47,619. The actual result is 50,912 and, as you can see, this yields a "decent" approximation.

- **428 = approximately 3000 / 7.** Therefore, if you need to calculate 678 x 440, you can round 440 to 428 and 428 to 3000 / 7.

- **666 = approximately 2000 / 3.** So, in the aforementioned 678 x 440, you can round 678 to 666 and 666 to 2000 / 3. This allows you to estimate the result as 2000 x 3000 / 3 / 7 = approximately 285,714. However, the actual result is 298,320.

- **999 = approximately 3000 / 3.**

- **1666 = approximately 10,000 / 6.**

Let's explore now the post-calculation technique known as "casting out nines," which is typically taught for multiplication but can actually be applied to all four basic operations to verify the accuracy of calculations.

Casting out nines for addition and subtraction

First, here's how the rule of nine can be applied for addition:

- For each addend, sum its digits. Then, sum the digits of the result until you have a single-digit number. We will call this operation "Modulus 9" or, to abbreviate, Mod9. This operation will be the "basis" of the technique for all four operations, and the reason it's called "Modulus 9" is that it corresponds to the remainder obtained when dividing that number by 9. For example, if you need to calculate 237 + 344, sum 2 + 3 + 7 = 12. Then sum 1 + 2 = 3. Now that you have obtained the Mod9 of the first addend, do the same with the second and sum 3 + 4 + 4 = 11. And 1 + 1 = 2.

- Now sum the Mod9 values of the addends, and if necessary, continue to add the digits of the result until you have a single-digit number. In our example, the final sums were 2 and 3. So, 2 + 3 = 5.

- Now take the result of your addition and apply Mod9. In this case, 237 + 344 equals 581. 5 + 8 + 1 = 14. 1 + 4 = 5.

Notice? The Mod9 value of the sum of the addends and the Mod9 value of the result *are the same*.

Importantly, this leads to a very significant consideration: if the two final sums do not match, you **can be sure** that you have made a mistake. However, if the two final figures are identical, you **can never be certain** that you have the correct result. In short, the rule of nine can help you realize if you have made a mistake, but it cannot give you absolute certainty that you have operated correctly.

The degree of uncertainty when verifying with casting out nines is quite low. It is rare for two final sums to match in an incorrect operation. More precisely, there is only an 11.11% probability that two randomly chosen numbers have the same Mod9. Also, note a fundamental property of casting out nines: in Mod9 calculations, **each 9 can be treated as a 0.**

But now let's move on to the subtraction check. In reality, the procedure is almost identical to that for addition, which is:

- For each operand, add up its digits. Then keep adding the digits of the result until you have a single-digit number. For example, if you need to calculate 788 - 215, add 7 + 8 + 8 = 23. Then add 2 + 3 = 5. Do the same with the other number by adding 2 + 1 + 5 = 8.

- Now, instead of adding the digits obtained in the previous step, subtract them in the same order imposed by the subtraction, and if you end up with a negative number, add 9. So, in this case you have 5 - 8 = -3. Adding 9 gives you 6.

- Now take the result of the subtraction and find its Mod9. In this case, you have 788 - 215 = 573 and 5 + 7 + 3 = 15. Additionally, adding 1 and 5 gives you 6, which again matches the first number obtained.
  Note: if the result of the subtraction had been negative, you would need to calculate the Mod9 as usual, which means adding the digits together until you reach a single-digit number.

However, you would then need to attach a minus sign to the modulus result and add 9.

- Let's consider another example to further clarify this last concept: suppose you need to verify the result of 45 - 388 = -343. The Mod9 of 45 is 9, and that of 388 is 1. Therefore, the first digit in the verification is 9 - 1 = 8.

- Now we find the Mod9 of the result: 3 + 4 + 3 = 10 and 1 + 0 = 1. This time we need to apply the negative sign and add 9, so 1 becomes -1 and by adding 9, we again get 8, which is the digit we were expecting.

Casting out nines for multiplication

This might be the most "well-known" and commonly used test, and it can be performed as follows:

- After completing your multiplication, draw an X on your writing tool.

- Place the Mod9 of the multiplicand on the left. For example, if you calculated 45 x 62, place 4 + 5 = 9.

- Place the Mod9 of the multiplier on the right. In the case of 45 x 62, 6 + 2 = 8.

- Place at the top the product of the just calculated Mod9s: 9 x 8 = 72. Here too, it's necessary to calculate the modulus until it is reduced to a single digit, so once again calculate 7 + 2 = 9.

- Now, place the Mod9 of the result of the original multiplication in the lower part. In this case, you have 45 x 62 = 2790. By adding 2 + 7 + 9 + 0, we can consider the 9 as a 0, and we immediately see that the result is 9.

- The check to be performed here is always the same: *the figure below must match the one above*. As usual, if they are the same, we've *likely* done well; otherwise, we have *certainly* made a mistake.

Casting out nines for division

The rule for division is identical to that for multiplication, with the necessary condition that it should only be applied when there are whole number results.

In particular, since division is the inverse operation of multiplication, meaning if a / b = c then c x b = a, we simply need to:

- Place to the left of the X the Mod9 of the result of the division as if it were the multiplicand of the inverse operation.
- Place the Mod9 of the divisor to the right of the X as if it were the multiplier of the inverse operation.
- Calculate the Mod9 of the product of these two moduli by summing the digits until reduced to a single digit.
- Place the Mod9 of the dividend below, as if it were the result of the inverse operation.
- Verify that the numbers match.

But let's consider an example to further clarify this concept, and assume we have calculated 308 / 44 = 7. This means that 7 x 44 = 308, and therefore:

- Let's place the Mod9 of 7 on the left (the result of the division and also the multiplier of the inverse), which is always 7.
- Let's place the Mod9 of 44 on the right (the divisor of the division as well as the multiplier of the inverse), which is 4 + 4 = 8.
- Let's take the product of these two numbers, so 7 x 8 = 56 and 5 + 6 = 11. Therefore, 1 + 1 = 2.
- Let's place the Mod9 of 308 (the dividend of the division as well as the result of the inverse) at the bottom: 3 + 0 + 8 = 11 and 1 + 1 = once again 2.

The speed and simplicity of these techniques can be an extraordinary value addition to your calculation abilities, especially if you find yourself performing arithmetic operations during exams, competitions, or tests, where you will certainly lack the time to verify the correctness of your results by recalculating everything from scratch.

Useful but probably not essential information: often alongside the rule of 9, there is a similar check called the "rule of 11." This method involves following exactly the same procedures as the rule of 9, but replacing the Modulus of 9 with "Modulus of 11," which involves finding the remainder of the division of the number by 11.

The calculation involves assigning weights to each digit and summing the products to compute the Mod11. For example, the Mod11 of 341 is calculated as $(3 \times 3) + (4 \times 2) + (1 \times 1) = 9 + 8 + 1 = 18$; then, 18 mod 11 equals 7.

The test with 11 is clearly a bit more complex due to the increased complexity of calculating Mod11, but it is slightly more reliable in terms of error checking, as there is a lower probability of ending up with the same modules in the presence of incorrect operations. Moreover, in cases of extreme uncertainty or when high precision is required, it is possible to use both types of tests simultaneously on the same operation and significantly increase the probability that it has been performed correctly (although unfortunately, it never reaches 100% certainty).

It is also worth concluding this chapter with a small clarification that, I hope, will also serve as an interesting curiosity for those who experience the technological world as mere "users": in the introduction, I mentioned the "substantial irrelevance of these practices in everyday life," given the way we increasingly rely on ever more precise and reliable tools to make our calculations. As a result, a tool for cross-verification like this one is, at least in theory, and except for what was discussed at the beginning of the chapter, becoming more obsolete. However, it is equally true that in all design fields where precision cannot be guaranteed due to physical limitations (consider data transmission through antennas that could be affected by excessive distances or interference), or is necessary for redundancy purposes (such as in the design of aircraft, bridges, medical instruments, etc.), similar cross-verification operations, like the much more complex checksum algorithms, are still fundamental for ensuring the accuracy of the data "navigating" our world. We could say, in short, that all these mathematical protocols are a true "silent superhero," constantly working to ensure that the systems governing our health, transportation, well-being, and even our human connections, are not victims of repeated, catastrophic failures.

The Historical Pill

The Second World War was fought as much on the battlefields as in radio rooms, where encrypted signals were transmitted, intercepted, decoded, and sent as diversions. It was by working on the "mathematics" necessary to decode these signals that each faction could predict and anticipate the opponent's moves, often significantly changing the war's balance. One of the most well-known cryptographic systems was generated by the Enigma machine, used by the German forces and deciphered by the Allies at the Bletchley Park group of mathematicians. It is suggested that the intelligence derived from the work at Bletchley Park "shortened the war by several years." It is uncertain how true this claim is, though it's perhaps virtually impossible to deny the significant impact of this "mathematical work" on the map of the contemporary world.

The Aphorism

"We are immersed in a web of complexity, where every part of the universe resonates with every other part. Mathematics is the language with which we manage to understand and navigate this web."
(Benoit Mandelbrot)

Mathematical Oddities

It's not unusual to find mathematical anecdotes, and even mathematical theorems, related to pizza. The "pizza theorem," for example, is a slightly counterintuitive result that suggests if you imagine cutting a pizza by making the first cut "at random" and then continue to make all other cuts from the same point while rotating by a fixed angle, you can continue to "fairly feed" diners, as you will always end up with equal pizza slices. Another "well-known trick" involves the "incredible" cost-effectiveness of always ordering a larger pizza when available. The example often given, although more applicable to large international chains than to Italian contexts, is that a 12" pizza is often "only slightly more expensive" than an 8" pizza. However, since this measurement refers to the diameter, and considering that the area of a circle is proportional to the square of the radius, the former has almost double the area of the latter. And this is a clear practical example of the "oddity" generated by the counterintuitive nature of non-linear size variations.

XIX - 10 strategies for calculating Squares... and one to save your life!

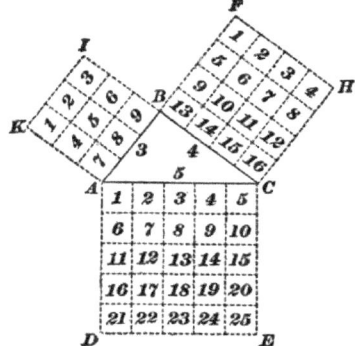

It is said that when Euler, the Enlightenment mathematician, worked on perfect squares, he was so captivated by their profound harmony that he came to believe they contained a proof of the existence of God. Clearly, the realm of metaphysical beliefs one might develop during the exploration of mathematics is beyond the scope of this book. Therefore, I will cautiously limit myself to observing that calculating the squares of whole numbers is often quite simple and, as seen in the chapter dedicated to "golden numbers," can lead to extraordinarily symmetrical and, I would dare say, beautiful results.

But let's move immediately to practice and review step-by-step all the strategies through which it is possible to perform these calculations optimally. While we may not attribute any divine implications to it, mastering these calculations can certainly simplify our lives. Especially since, as we will see through a case at the end of the chapter (and have surely already glimpsed in the "larger pizza trick" from a few pages back), mastering this type of calculation can help us navigate the pitfalls of all those phenomena that grow more rapidly than linear ones (consider something that is worth 1 on day 1, 10 on day 10, and so on). That's indeed the case for variables following geometric (like in the case of quadratic growth: something is worth 1 on day 1, 100 on day 10, etc.) or exponential growth (such as any positive integer that grows "incredibly fast" as we increase its exponent). Our difficulty in perceiving and understanding exponential or geometric phenomena can be attributed to various "biases" and cognitive limitations, likely because linear relationships, besides being more intuitive, can easily be associated with various common phenomena, like the growth of plants or the passage of time. Anything that exceeds, in acceleration, a linear growth ultimately challenges and surpasses our capacity for prediction and anticipation. It literally makes us "wake up one day" to find that something we thought was far off "has already happened." Consider how compound interest works in the world of finance: capital growing annually at "relatively" small interest rates suddenly finds itself doubled; or how a few infected individuals can quickly lead to a mass epidemic (a notion that sadly will sound familiar to many). Or the more encouraging case of the "singularity theory," a theoretical vision of the future where technological progress, by essentially "feeding on itself," can accelerate so rapidly that machines begin advancing at speeds no longer comprehensible to humans. Think also of all the "anecdotes of exponential growth" that challenge our estimation skills to the limit: for instance, the legend of the inventor of chess mentioned a few pages back; or the one stating that you "only" need to fold a piece of paper 23 times for it to reach the height of Everest and 42 times for it to cover the distance between Earth and the Sun. But let's come back down to earth for a moment and analyze our techniques for quickly calculating squares, hoping that they will also provide us with a good set of cognitive tools. These tools can help us comprehend fast-evolving phenomena and apply similar strategies to areas we wish to grow, like our capital or knowledge.

1 - Learning Basic Squares

Memorizing the squares of the 20 smallest positive integers can be incredibly useful for speeding up the calculation of squares of larger numbers. That's why it can be a good idea to start the chapter with this specific table. Repeat it, try to learn it "by heart," and despite the possible initial tedium, you'll find that many of the calculations demonstrated in the subsequent techniques will become much more straightforward thanks to the groundwork you've done:

$$1^2 = 1$$
$$2^2 = 4$$
$$3^2 = 9$$
$$4^2 = 16$$
$$5^2 = 25$$
$$6^2 = 36$$
$$7^2 = 49$$
$$8^2 = 64$$
$$9^2 = 81$$
$$10^2 = 100$$
$$11^2 = 121$$
$$12^2 = 144$$
$$13^2 = 169$$
$$14^2 = 196$$
$$15^2 = 225$$
$$16^2 = 256$$
$$17^2 = 289$$
$$18^2 = 324$$
$$19^2 = 361$$
$$20^2 = 400$$

2 - Exploit the technique of multiplying numbers between 11 and 19

Yes, of course, sometimes it isn't exactly simple to memorize the squares of the first 20 positive integers. Let's say that most people typically manage to remember with reasonable ease the first 10 (also because they appear in multiplication tables) and at most those of 11, 12, and 13. For the others, however, we can rely on a calculation technique inspired by a multiplication strategy we saw a few pages ago, specifically for numbers between 11 and 19, as seen in Chapter XIII. By adapting it to two identical numbers, we find that to "square" numbers in this category, we can use this simple strategy:

1. Sum the number to its units digit.

2. Multiply the result by 10.

3. Add to the result in point 2 the square of the units.

The square of 17, for example, will be:

- 17 + its units 7 = 24
- 24 x 10 = 240
- 240 + (7^2 = 49) = 289

It remains true, of course, that knowing these squares "by heart" provides greater speed compared to having to recalculate them each time. Moreover, this technique can clearly be used to calculate the squares of numbers like 110, 120, 130, etc. The key point, as is extremely easy to understand, will be to add two trailing zeros to the final result.

3 - Strategy for numbers ending in "1"

Before moving on to further "general" strategies, let's introduce some tricks to quickly calculate the square of numbers ending in "1." Here's a method:

1. Calculate the square of the number without its unit digit and place this result on the left.

2. Double the number without its unit digit and place this doubled value to the right of the previous result.

3. Finally, append 1 to the end.

Example for 41^2:

1. Calculate $4^2 = 16$. Place it on the left.

2. Double 4 to get 8. Form a new number 168 by placing 8 to the right of 16.

3. Append 1, resulting in 1681.

Example for 431^2:

1. Calculate 43^2. Use any preferred technique; the result is 1849.

2. Double 43 to get 86. Form a new partial number 184986 by placing 86 to the right of 1849.

3. Append 1, leading to 1849861.

2. Mathematical Errors: Revised steps and examples to ensure correctness in the method and outcome.

4 - Strategy for numbers ending in "5"

This is one of the most well-known and appreciated quadrature strategies, given the incredible simplicity and speed of calculation it offers.

In particular, if you want to square a number that ends in 5, all you have to do is:

1. Multiply the number formed by removing the units digit by the next consecutive number.

2. Place a 25 to the end of the result.

Let's start with a simple example: if we wanted to calculate 75^2, we would need to:

- Calculate 7 x 8 = 56.
- Append 25 = 5625.

If we have a three-digit number, once again, the method doesn't change even though it becomes slightly more "costly." In fact, if we wanted to calculate 335^2, for example, we would have to perform:

- 33 x 34 = 1122 (and here you can use any of the multiplication techniques explained in the previous chapters).
- Append 25 = 112,225

5 - Strategy for numbers ending in "25"

Squaring numbers that end in "25" is also very simple to calculate. Specifically, you need to:

1. Square the number without the final "25."
2. Add to this square half of the number without the final "25."
3. Multiply it by 10.
4. Append 625.

Let's suppose, for example, that we want to calculate 825^2. We would need to follow these simple steps:

- 8 times 8 equals 64.
- 64 + (half of 8) 4 = 68.
- 68 multiplied by 10 equals 680.
- Appending 625 gives us the result, which is 680,625.

6 - Strategy for numbers starting with "5"

This is another calculation strategy that is extremely easy to implement. Let's begin exploring it by first seeing how to square a number between 51 and 59:

1. Calculate the square of the units digit of the number and place it on the right.

2. Now take the number to be squared, subtract 25 from it, and place the result on the left.

So, if you wanted to calculate 56^2, you would:

- Calculate $6^2 = 36$.
- Calculate $56 - 25 = 31$.
- The final result is exactly 3136.

The strategy, however, is not limited to two-digit numbers but can be extended to any number starting with "5," with the only distinctions being that one must:

- Calculate the square of everything that comes after the 5 and place it on the right.
- Take 25 and add zeros until you have a number with the same number of digits as the number to be squared (so you should get 250 if you have 553, 2500 if you have 5811, etc.). Then, subtract it from the number to be squared and place the result on the left.

Thus, the calculation, for example, of 531^2 will consist of the following steps:

- Calculate $31^2 = 961$. Place it on the right.
- Calculate $531 - 250 = 281$. Place it on the left.
- The final result is exactly 281,961.

7 - Reuse the technique for multiplying numbers "equally distant from an integer."

In Chapter XIII, we saw that it is possible to multiply two numbers by calculating the square of their mean minus the square of their distance from the mean.

Now, therefore, if we simply reverse the formula, we find that it is possible to calculate the square of a number by multiplying together two numbers that have the original number as their mean and then adding the square of their distance from the original number.

But how can we find two numbers whose average is the one we want to calculate the square of? In reality, it's quite simple, as you just choose the preceding and the following number, or those obtained by adding +2 and -2, +3 and -3, +4 and -4, etc.

In general, the best rule to simplify this method is to find two numbers such that the resulting multiplication is very simple. This can be achieved, for example, by "reaching" a multiple of 10 or by using any of the rules that make multiplication "simple" as discussed in Chapter XIII. Choosing to minimize the distance can also simplify calculations, but this is certainly a secondary consideration compared to the other.

Let's say, for example, we want to calculate 38^2: we could:

- **Multiply 37 by 39 and add 1** (distance = 1, so the square of the distance = 1).

- **Multiply 36 by 40 and add 4** (distance = 2 so the square of the distance = 4).

- **Multiply 35 by 41 and add 9** (distance = 3, so the square of the distance = 9).

- ... and so on.

In this case, the simplest approach is definitely to calculate 36 x 40 = 1440 and then add 4, thus obtaining 1444, which is precisely the square we were looking for. Notice how this technique allows you to calculate squares in a really quick and simple way, reducing the calculation to a trivial multiplication plus a trivial addition.

8 - Take advantage of the proximity to another square

By using a variation of the formula seen in the previous section, we can easily calculate the square of a number "n" if we know (or can easily calculate) the square of another number "r" that is "close" to it. Specifically, we should:

- Add n and r.

- Multiply this sum by the distance between n and r (remembering that the distance is obtained by subtracting the smaller from the larger number).

- Take the square of r, which we already know (or have managed to calculate easily), and add the result of the previous product if the reference r is less than n. However, if r is greater than n, we will need to subtract that result from the "known" square.

Here, as a reference, you can use a multiple of 10 or a number that begins or ends with 1 or 5, given the ease of calculation these categories of numbers provide. But let's get straight to an example by trying to calculate 58^2:

- We can use the number 55 as a reference. Its square, which can be immediately calculated using the strategy for numbers that end (or begin) with "5," is 3025.

- We add 55 and 58 to get 113.

- We multiply 113 by 3, which is the distance between 55 and 58, resulting in 339.

- Now all we have to do is "understand" whether this 339 should be added to or subtracted from the initial 3025. So, let's check if the chosen reference is greater or lesser, and noticing that 55 is less than 58, we should then add: 3025 + 339 = 3364, which is indeed the square of 58.

9 - Reuse mental multiplication techniques for two and three-digit numbers.

The multiplication technique for two and three-digit numbers seen in previous chapters can also aid us in calculating squares. Just like the multiplication technique for numbers from 11 to 19, we simply need to adapt it to two identical numbers and follow the exact same procedures. Specifically, in these cases, we can:

- Calculate the square of the tens digit and place it on the left in the partial result.
- Calculate the product of the units and tens, double it, and place it in the center of the partial result.
- Calculate the square of the units and place it to the right in the partial result.
- Perform all the necessary carry-overs.

So, if we wanted to calculate 87^2, we would need to:

- Calculate $8^2 = 64$, and place it on the left.
- Calculate the double product of $7 \times 8 = 56$, which multiplied by 2 is 112. The partial result is 64_112.
- Now calculate $7^2 = 49$. The result will be 64_112_49.
- By carrying over, we have:

$$64\_116\_9$$

$$75\_6\_9$$

And 7569 is the number we were looking for.

For three-digit numbers, the procedure, clearly, becomes slightly more complicated, but it's still possible to use the same method in a relatively efficient manner:

Procedure 1

- Square the number formed by the first two digits and place it on the left. So, for example, in the case of squaring 458, you first calculate the 45^2, which is 2025 (and here you have an advantage because 45 can be calculated immediately).
- Now calculate the product of the first two digits with the units, double it, and place it in the center. So in this example, you need to calculate 45 x 8 x 2 = 720. The partial result is 2025_720.
- Finally, calculate the square of the units and place it to the right. In this case, you have 8 x 8 = 64, and the partial result is 2025_720_64.
- Calculate the carryovers:
  2025_726_4
  2097_6_4
  And indeed, 209,764 was the number we were looking for.

Procedure 2

- Note that this is the "cross" procedure seen in the graphical multiplication of Chapter XV, but starting from left to right. Considering that the carrying is done at the end, the order of procedure is not important at all.
- Calculate the square of the hundreds digit and place it on the left. For example, if you want to calculate the square of 612, start by squaring 6 = 36.
- Multiply the hundreds by the tens, double it, and place it to the right of the previous number. So, in this case, you need to calculate 6 x 1 x 2 = 12. The partial result is 36_12.
- Multiply now hundreds by units, double it, and add the square of the tens. In the example you have: 6 x 2 x 2 + 1 x 1 = 25. The partial result is 36_12_25.
- Multiply tens by units, double it, and place it to the right of the previous number. Here you have 1 x 2 x 2 = 4. So you end up with 36_12_25_4.

- Square the units: 1296_144_625_16_16.
- Carry out all the calculations:
  36_14_5_4_4
  37_4_5_4_4
  And indeed, 374,544 is exactly the square of 612.

This "distinction" between the two procedures is also useful for working with larger numbers. In fact, even for numbers with 4, 5, or more digits, you can group the first few digits and refer to the first type of procedure (which, for example, in the case of 4-digit numbers would be "First three digits squared" + "First three digits multiplied by units doubled" + "Units by units"), or choose to calculate digit by digit and refer to the second type (the one seen in Trachtenberg's graphical multiplication).

It's up to you to understand what seems easiest and, if possible, to identify among the various numbers any known products that might facilitate your calculations.

10 - Reuse the multiplication technique for numbers close to a power of 10

The technique we will explore in this section can also be extremely useful for calculating the squares of very large numbers, as it allows us to reduce them to the calculation of very small squares. For these, any of the previous techniques can then be applied.

Drawing inspiration from the technique of multiplying numbers with the same characteristic, we can distinguish between two "sub-cases":

Numbers slightly smaller than a power of 10

- Keep in mind that if you are "sufficiently close" to the next highest power of 10, the square should have a number of digits equal to twice those of the number. For example, 96^2 should have four digits.

- Calculate the complement of the number to the next higher power of 10. If it's not immediate, apply the Sutra "All from 9, the last from 10." For 96 → 4.
- Calculate the square of that complement and place it on the right in the result. For 96 → 4 → __16.
- Now subtract from the initial number its complement with respect to the power of 10 and place it on the left in the result. 96 - 4 = 92.
- If the number of digits is less than what was calculated in the first step, you will need to add some "zeros" in the middle. In the case of 96, it won't be necessary, and in fact, the result is 9216.

But let's immediately try another example, attempting to calculate 9987^2:

- The result should have 8 digits, since 4 x 2 = 8.
- 9987 complement to 10,000 is 13.
- $13^2 = 169$. We will place it on the right.
- We subtract 13 from 9987, obtaining 9974. We place it on the left.
- So far we hence have 9974_169. We said the result should have 8 digits, so we can insert a zero in the middle, obtaining the final result, which is 99,740,169.

As already noted in the case of multiplication, this technique allows operations with even enormous numbers at extraordinary speeds. However, the further one moves away from powers of 10, the more things become complicated. This is because, despite the procedure remaining valid, the complements grow larger, and consequently, calculating their square gradually becomes longer and more complex, eventually matching the complexity of the original number.

Numbers slightly larger than a power of 10

- Keep in mind that if you are sufficiently close to the lower power of 10, the result should have a number of digits equal to twice the number of digits of the number, minus one.
- Calculate the complement of the number to the nearest power of 10 above it and square it. Then, place that square to the right in the result.
- Add the number to its complement and place it on the left in the result.
- If the number of digits is greater than calculated in the first step, you must take the leftmost digit of the right part and add it to the left.

But let's clarify the procedure once and for all with an example, assuming we want to calculate 1094^2:

- The result should have 7 digits, that is, twice 4 minus one.
- The complement to 1000 is obviously 94. I can easily calculate 94^2 using the method discussed in the previous section, which involves placing the square of the complement to 100 on the right (i.e., 6 x 6 = 36) and the number minus its complement (94 - 6 = 88) on the left. Result = 8836, to be put on the right in this result.
- Now add 94 to 1094, obtaining 1188, to be placed on the left in the result.
- The partial result now is 1188_8836. However, we mentioned that the result should have 7 digits instead of 8, which is why we take the leftmost digit of the right part (the last "8" of 8836) and add it to 1188, thus finally obtaining 1196_836 = 1,196,836.

The techniques for calculating squares end here, and now that you hopefully have at least a partial mastery of them, here's an interesting fact: If you calculate the square of 15, using any of the aforementioned techniques, you'll find that the result is 225. Now, calculate the square of 25, which is a "slight increase" from 15, and you'll get 625, which is nearly three times the previous square.

Now that we're close to the end, let's conclude the subject by trying to extract a basic concept from kinematic physics: kinetic energy, which is the energy of a body associated with its motion, increases proportionally with the square of its velocity. Just above, you calculated that the energy associated with a car traveling at *90 km/h* (25 m/s) is practically triple that of a car going at *54 km/h* (15 m/s). A relatively small increase in speed, but the energy is tripled, and the damages in the event of an accident are tripled, if not much greater. A "simple" case of geometric growth (and our "easy underestimation" of it) that explains in the clearest way possible why we should not speed on the road: because even small increases in speed will greatly worsen the damages in case of an accident, just as, conversely, keeping the speed low will reduce any risk for you and your passengers by several orders of magnitude.

The Exercise

Let's briefly tackle the topic of "cubes" (and, for those who forgot, it means raising a number to the power of three). As far as we know, there aren't quick and easy everyday strategies for calculating cubes of whole numbers or otherwise. However, since the cube of a number "n" is the same as "the number multiplied by itself twice," that is, n x n x n, or n squared times n, what "quick" multiplication or squaring strategy could you use to calculate, for example, the cube of 11, 25, or 69?

The Historical Pill

In 1931, mathematician Kurt Gödel demonstrated that in any sufficiently complex system of mathematical rules, there will always be some statements that cannot be proven either true or false using only the rules of that system. For example, the moment we say "this statement is false," we find ourselves in a paradox: if the statement is true, then it must be false as it claims, but if it is false, then it is actually true. This is an example of a statement that cannot be definitively resolved within our system of linguistic and logical rules. Gödel showed that the same happens with any conceivable mathematical system: it will either be incomplete, or any attempt to "complete" it will lead to new paradoxes and inconsistencies. This result, formalized in the so-called "Gödel's Incompleteness Theorems," shocked the world of mathematics by essentially proving that there are limits to what we can know, and some mysteries might remain unsolved forever. This, indeed, is another one of those answers that can easily leave us with a real sense of "existential void."

The Aphorism

"The problem with our societies is not that we are irrational, but that we are rational in too simplistic a way. We need a more complex rationality, one that acknowledges the limits of pure logic and embraces uncertainty. This is the fundamental teaching of mathematics."
(Nicholas Nassim Taleb)

Mathematical Oddities

But let's return to the paradoxes of infinity, and in particular, to the so-called "Galileo's Paradox." Consider the whole numbers (1, 2, 3, 4, ...) and the even numbers (2, 4, 6, 8, ...). Intuitively, one might easily say that there should be more whole numbers than even numbers because the even numbers are just a part of the whole numbers. However, there is always a one-to-one correspondence between the whole numbers and the even numbers (1 corresponds to 2, 2 corresponds to 4, 3 corresponds to 6, and so on), demonstrating that there are as many even numbers as there are whole numbers. But how is this possible if one group is a subset of the other?

XX - Square root, cube root, and... thirteenth root!

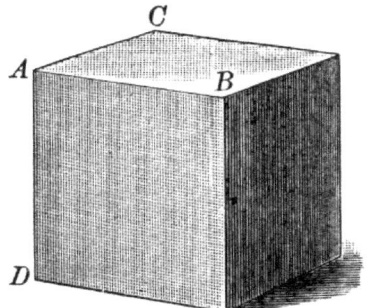

The extraction of the square root of a non-perfect square (meaning its root is not an integer) can often prove to be a rather cumbersome and frustrating operation. This is why, as good rapid calculators always in search of a quicker and more effective strategic path to complete our tasks, we will once again resort to the best strategies to avoid the most tedious procedures, aiming to achieve the best result in the shortest possible time. A historical note of interest, hopefully: it is said that when the Pythagoreans "discovered" that $\sqrt{2}$ ($\approx$ 1.414213 etc.) was a number that "did not terminate" (having infinite decimal digits as the calculation continued), and at the same time did not repeat (it was not periodic like an "ordinary" $1/3 = 0.3333$ etc.), and could not be expressed through any kind of fractions, their beliefs were completely "shattered"; it was, apparently, the first discovery of irrational numbers, numbers that cannot be expressed through division. Such an event, like every time new "lenses" are added to view the world, is capable of providing means to build new theories, concepts, and practical tools. Consider even how $\sqrt{2}$ is an integral part of the design of the seemingly "mundane" international paper sizes (such as A4, A3, etc.): the 1:$\sqrt{2}$ ratio ensures that when a sheet is divided in half, the width-to-height ratio is preserved.

How to identify a non-perfect square.

First of all, let's try to forget for a moment about the irrationality of certain constants, and look at a "trick" that can help us immediately identify a number that is not a perfect square: *a perfect square always ends in 1, 4, 5, 6, 9, or 0. The reverse, of course, is not always true, and hence not all numbers ending in 1, 4, 5, 6, 9 or 0 are perfect squares.* Thanks to this rule, when we see a number that doesn't end with one of these digits, we can *at least* be sure that its root will not be an integer.

Anyway, in theory we also have a technique that allows us to identify a perfect square immediately. Let's say that, since it is not essential for the calculation strategies we will see in this chapter, it's not worth delving too deeply into this topic. Let's simply introduce it by stating that *a number is a perfect square when, if broken down into prime factors, all of its factors have an even exponent* (or appear an even number of times, which is the same thing). I'll leave all the extra proofs and deeper explanations with you and move on.

Two methods to quickly approximate Square Root

A preliminary technique for approximating the calculation of a square root involves utilizing a mathematical relationship between the number "a" whose root we wish to calculate and a number "b" whose root we already know (which should be easier for us if we have properly applied the criteria explained in the previous chapter). Specifically, we have that $\sqrt{a}$ (the unknown root) is approximately equal to:

$$\sqrt{b} \text{ (which is the known root)} + ((a - b) / 2 \times \sqrt{b})$$

where the approximation is increasingly closer to the real result as "a" and "b" are closer to each other. However, since formulas without examples might only cause confusion, let's immediately demonstrate a practical application of the technique and attempt to calculate the square root of 5. We naturally take 4, We naturally take 4, whose square root we already know, as the reference number, as the reference number and calculate:

- $\sqrt{4} = 2$, which is our known root.
- We calculate: $(a - b) / (2 \times \sqrt{b}) = (5 - 4) / (2 \times 2) = 1 / 4 = 0.25$
- Then add this result to the known root: $2 + 0.25 = 2.25$.
- We have thus found that the approximate square root of 5 is 2.25. The actual value is approximately 2.236, but as you can see, the approximation we've obtained is "quite good."

If we want greater precision, we must use a second method of incremental approximation (that is, *the more steps we carry out, the better the approximation will be*), which works as follows:

- We take the number for which we want to find the square root and then select the perfect square immediately below it (for example, if we want the square root of 38, we take 36).
- Now we calculate the average between our number and the perfect square in question (for example, in this case, $(38 + 36) / 2 = 37$) and divide the number (our original number) by the initial approximation (thus we will perform $38 / 6 = 6.3333$). This is an initial approximation of the root.
- But if we want to achieve an even better approximation, we can take the original number from which we are extracting the root (still 38) and divide it by the approximation obtained in the previous step. In our example, this means dividing 38 by 6.3333, resulting in approximately 6.0005.
- Final level, final optional step, best possible approximation: take the number from the previous step and calculate the average with the one obtained in step 2. The result, rounded to four decimal places, is 6.1644. The real root of 38, when rounded to four decimal places, is approximately 6.1644. As you can see, everything depends on the number of steps we want to take, but in general, the degree of precision of this formula is truly exceptional.

Tricks for Cube Root and Fifth Root

As the root degree increases, the associated operations become more complex and much less useful in everyday life. Therefore, let's try to stick closely to the realm of fun and easily applicable "math tricks."

Extracting the cube root of a number is not a simple operation, which is why in this section we will discuss only a trick for quickly extracting the roots of perfect cubes with 4, 5, or 6 digits.

The first step in mastering this trick is to learn the cubes from 1 to 10:

$$1^3 = 1$$
$$2^3 = 8$$
$$3^3 = 27$$
$$4^3 = 64$$
$$5^3 = 125$$
$$6^3 = 216$$
$$7^3 = 343$$
$$8^3 = 512$$
$$9^3 = 729.$$
$$10^3 = 1000$$

Now, after having possibly simplified memorization with a bit of phonetic conversion, remember that, unlike what happens with the square root, the last digit of a perfect cube immediately determines what the last digit of its root is. More specifically:

- If the cube ends with 0, its root ends with 0.
- If the cube ends in 1, its root ends in 1.
- If the cube ends with 2, its root ends with 8.
- If the cube ends with 3, its root ends with 7.

- If the cube ends with 4, its root ends with 4.
- If the cube ends in 5, its root ends in 5.
- If the cube ends with 6, its root ends with 6.
- If the cube ends with 7, its root ends with 3.
- If the cube ends with 8, its root ends with 2.
- If the cube ends in 9, its root ends in 9.

And this table can be easily memorized by remembering that the digits of the cube and the root almost always correspond, except when the cube ends with 2, 3, 7, and 8. In such cases, the root shows the digit's complement to 10 instead of the digit itself.

Given these two tables, you have everything you need to calculate the root of your perfect cube of 4, 5, or 6 digits. So:

- Its root will have 2 digits.
- The units digit will be provided by the second table.
- To find the tens digit, remove the last three digits from your perfect cube. Compare what's left to a table of perfect cubes. As you go higher in the table, you'll notice smaller cubes that are less than the remaining number but getting closer to it. Keep going until you find a cube that is larger. This step helps identify the tens digit.
- Then take the largest perfect cube that is still less than your number: its root (cubic, of course) will be the tens digit you were looking for.

Let's immediately use an example to clearly establish the concept, and let's say we want to extract the cube root of 85,184. The unit digit is found immediately, and it's 4. For the tens digit, we need to remove the last three digits of the number, leaving us with 85, and then check in the table for the largest perfect cube less than 85. As we go up, 1 is smaller, 8 is smaller, 27 is smaller, 64 is smaller, but 125 is larger. Therefore, we stop at 64: its cube root is 4, which is our tens digit. Indeed, the cube root of 85,184 is precisely 44.

The last trick I want to teach you about cube roots is one that allows you to approximate it using any non-scientific calculator (or calculator app) equipped only with the square root function. Here's what you need to do:

- Enter the number from which you want to extract the root.
- Press the square root key once.
- Press the "multiply" button.
- Press the square root key twice.
- Press the "multiply" button.
- Press the square root button four times.
- Press the "multiply" button.
- Press the square root button eight times.
- Press the "multiply" button.
- ...

And you must continue, doubling the number of times you press the "square root" button until the result no longer changes significantly after pressing the "multiplication" button. At that point, you need to press the square root button one last time, and you'll have the cube root displayed.

Curiosity: want to approximate the calculation of the fifth root instead of the cube root? Replace the "multiplication" key with the "division" key!

Thirteenth Root?

I don't believe any of you will ever have to deal with the thirteenth root of a number in your life, and that's why we'll conclude this chapter with a few simple anecdotal curiosities.

The extraction of the thirteenth root is quite a famous practice in the field of world records for fast mental arithmetic. The very first record was held by Herbert B. De Grote, who calculated the thirteenth root of a 100-digit number in 23 minutes. However, nowadays, the German mental calculator Gert Mittring has managed to calculate it in 11.8 seconds.

Clearly, Mittring, like others who have attempted to break the record before him, performs his calculations using a precise technique of considerable complexity, which is only partially known. Additionally, it is assumed that the number being worked on has an exact thirteenth root.

In any case, without wanting to delve into the details of such a calculation technique, we can say, purely for informational purposes, and always with the premise that we are discussing entire thirteenth roots, that:

- The thirteenth root of a number has the same last digit as the number itself.

- The thirteenth root of a 100-digit number is definitely 8 digits long and is certainly between 41,246,264 and 49,238,826. For this reason, it is generally considered a "fairly simple" operation, based on calculating a logarithm, some multiplications, and some exponentiation. No big deal!

- The thirteenth root of a 200-digit number is definitely 16 digits long and lies between 2,030,917,620,904,736 and 2,424,462,017,082,328. This, however, is considered a "fairly complex" operation. The world record in this regard (at least at the time of writing this book) is about 5 minutes. One might rightly argue that it's a relatively short calculation time given the complexity of the operation, but it is still about 30 times longer than what the fastest minds in the world take to calculate a 100-digit root.

Some individuals have speculated about associating profound mystical and esoteric meanings with a similar operation. Everything seems to be linked to the "mythology of 13," a number associated with countless symbolic meanings: consider its iconographic connection with Lucifer's rebellion against the heavens, the fact that it is the number of guests at the "Last Supper" attended by Jesus Christ, or the fact that the "Death" card in tarot is indeed the thirteenth. But aside from giving too much credit to the "metaphysical significance" attributed to certain numbers, my hope is, once again, to provide further, tangible evidence of how even something seemingly "cold" and "soulless" like numbers have been capable, since the dawn of time, of sparking the wildest imaginations within us.

The Historical Pill

As many may know, there are seven mathematical problems known as the "Millennium Prize Problems" that, if solved, guarantee a million-dollar prize due to their implications in fields like physics and cryptography. One of these, known as the "Poincaré Conjecture," has been solved, but the prize winner, a Russian mathematician named Grigori Perelman, declined the money. There's also the possibility that solutions for the remaining six problems might not even exist; however, if by any chance a solution for any of them were to be published in a sufficient number of peer-reviewed scientific journals, it could lay the groundwork for significant advancements not only in mathematical research, but in the entire scientific landscape. Solving the problem related to the "Riemann Hypothesis," for example, could revolutionize the world of cryptography; or, reaching a solution for the so-called "P vs. NP" problem, we could solve many extremely complex problems much more efficiently, such as optimizing logistics routes, predicting the proteins that a gene codes for, or even solving some of the toughest problems in the field of artificial intelligence.

The Aphorism

"Archimedes will be remembered when Aeschylus is long forgotten, for languages die but mathematical ideas are eternal."
(G. H. Hardy)

Mathematical Oddities

If the universe is truly infinite and every single event, regardless of its complexity, has the possibility of occurring, then it can be deduced that the existence of an exact copy of an individual somewhere else in the universe is not just possible, but highly probable. On an even deeper level, the infinity of the universe might imply the existence of an infinite number of exact replicas of an individual. It is clear, however, that the physical reality of these facts might be "much more complex and at the same time less fascinating than this," but considering this paradox and all those of past "oddities," I still find it extremely poetic how mathematics represents a true "privileged gateway" through which to connect our own limitations to the unlimited and the absolute.

- *The Speed Math Bible* -

XXI - Mathematical tools

Given the density and complexity of some of the previous chapters, in this final chapter we will venture into a "quick series" of many small tools and mathematical formulas for everyday life. Hopefully, by the end of the reading, these will provide us the opportunity to "exercise our neurons" sufficiently, manage some everyday situations more effectively, and thus increase our own level of connection and "intimacy" with the mathematical universe and its infinite facets.

Estimate, with Gauss

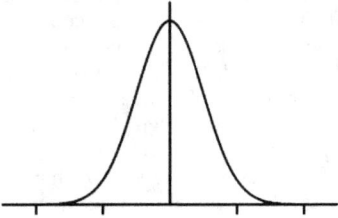

Imagine you finally want to clean your house after four weeks of "wild living." What might be the best way to estimate the time needed for this task, allowing you to organize the rest of your day effectively? Well, here is how we could make our estimate through an extraordinarily elegant mathematical procedure:

1. First, think positively. How long would it take in the best case scenario? Let's call it the optimistic estimate (O), for example, 1 hour.

2. Then, consider the worst-case scenario. How long would it take in the worst-case scenario? Here's the pessimistic estimate (P), for example 5 hours.

3. Finally, be realistic. How much time would you most likely take? This is the most probable estimate (M), for example, 2 hours.

4. Now, calculate the estimate using this formula: (O + 4M + P) / 6. So, in this case, $(1 + 4*2 + 5) / 6 \approx 2.33$ hours.

But why do we use this method of weighted average, known as "PERT estimation"? And why not simply use the mathematical average between the best and worst case, or even just the "realistic" case? The answer lies in the fact that, as mentioned a few pages back, the Gaussian curve, or normal distribution, is a mathematical model that often very well describes how the uncertainty of certain events is distributed in phenomena of all kinds, whether natural or artificial. Look at the image at the beginning of the section: the height of each point shows the probability with which something occurs, very high in the middle, that is, in the "average" cases of the considered phenomenon, and increasingly lower as one approaches the extremes. So if, for example, the curve represented a distribution of height within a population, it would suggest that there are many people "in the average," and few very short or very tall. If it described the outcome of "many" rolls of two dice, in the center would be the probability of rolling a 7 (the most likely case, as it can be obtained with the most different combinations), at the extremes, the two rarest cases (12 and 2), while all other results would be "in between."

The evaluation of such an average is therefore "slightly more sophisticated" than a normal arithmetic mean, as it duly considers the weights of all possible cases depending on their "possible occurrence." For this reason, it is a practice widely used in professional Project Management to assess estimates for business projects, whether large or small. But not only that: you can use it to estimate, for example, the cost of a project by considering the best-case scenario (everything goes smoothly and suppliers offer discounts), the worst-case scenario (unexpected problems arise and costs increase), and the most likely scenario (some minor setbacks but nothing too serious). Or, again, to calculate the amount of food to prepare for a party, the number of books you will be able to read in a year, and so on.

When to say "Yes"?

A very interesting mathematical problem is the so-called "optimal stopping," which can be described as follows: there is a deck of cards face down on a table. A number is printed on each card, ranging from 0 to an unknown maximum. Now, we can pick one card at a time, look at it, and decide each time whether to keep it or discard it; all with the aim of trying to keep the card with the highest possible number and thus to avoid the risk of passing on potentially high-scoring cards. This leads us to a specific question: "How large should the sample of cards be that we draw to be reasonably sure of having picked a card with one of the highest scores?"

This problem, given its applicability in countless practical fields, has always stimulated our "mathematical imagination." For instance, a mathematician, Charles Brenner, reintroduced it by applying it to the context of searching for one's ideal partner, comparing the deck of cards to one's "search period," and the score to the level of connection one is able to establish with a potential partner encountered.

The solution to this problem has been well known for a long time, and the optimal sample from which it's "wise to stop" after drawing at least one "acceptable" element is equal to **1/e**, where **e = Euler's number ≈ 2.718**, so its inverse is ≈ **0.368**. That is, approximately, we should have developed a "reasonably" accurate idea of the distribution of the elements in the deck after examining roughly 37% (slightly more than a third) of the total units; from there, it makes sense to stop at the first "better than average" element encountered.

Returning to Brenner's example: assuming, for the sake of simplicity, that the "optimal search period" for a man spans 20 to 45, the "optimal point" occurs around age 29. At that age, it should typically become advantageous for a man to choose, as a long-term partner, the first person with whom he feels a stronger connection than with those met before.

Clearly, as the more "romantic" among you might already be objecting, love is not something you choose, and it cannot be calculated with numbers. However, I still find it extremely interesting to understand how the dilemmas, theorems, and premises of mathematics allow us to better comprehend (and thus theoretically better manage) countless practical situations. This formula, in fact, could be used to make more conscious and informed decisions in all those circumstances where we have to "sooner or later" decide whether to say yes or "hope for something better first": consider buying a house (choose the one that seems the best after an "observation period" equal to 1/e of the total period), the timing of investing in a business, starting a family, and so on.

Ideal weight (or almost)

There are various formulas to calculate a person's ideal weight, starting from the now outdated (as it is quite "unreliable" and inaccurate) method of subtracting 100 from one's height in centimeters, to more complex ones that consider a person's body fat index. However, one of the simplest (as it doesn't require measurements beyond height) and more "acceptable" methods (though it remains theoretical and approximate, as it doesn't take specific cases into account; therefore, "take it with a grain of salt") is the so-called "Keys Formula," which states that the ideal weight is equal to:

(Height in meters squared) × 22.1 for males

(Height in meters squared) × 20.6 for females

Hypothetical Close Encounters

For all those who may not already know, there is a mathematical equation known as the "Drake Equation," named after the American scientist Frank Drake, which is supposed to provide us with a "reasonably accurate" number of extraterrestrial civilizations with which communication should be possible in the universe. In fact, after multiplying:

- The rate of star formation in the galaxy.
- The fraction of stars that have planets.
- The average number of habitable planets for each star with planets.
- The fraction of habitable planets on which life could have actually developed.
- The fraction of planets on which such life could have become intelligent.
- The fraction of planets where such intelligent life might have developed the capability (and the willingness) to communicate with other forms of life.

Based on some astronomical hypotheses and conjectures, Frank Drake originally suggested that this equation might yield a number *around 10*, but this is really a speculative estimate rather than a definitive result. However, as often pointed out by many, the apparent contradiction presented by the "Fermi Paradox" arises: if extraterrestrial civilizations exist, why haven't we encountered any signs of them? This paradox underscores the need for further reflection on potential technological or temporal constraints in extraterrestrial communication. We must consider that the development of interstellar signal transmission technologies might be hindered by the laws of physics. Moreover, the evolution of a civilization might not necessarily lead to advanced technologies or an interest in external communication.

Cosmic Card Decks

If you have the opportunity to combine the 7 classic notes of the musical scale and want to understand how many different 12-note melodies you can compose using the 7 notes freely at your disposal, you simply need to raise 7 (the number of notes to choose from) to the 12th power (the melody length). The result is that, even using a simple children's keyboard for ages 2 to 3, you can compose as many as 13,841,287,201 possible melodies, each distinct from the other by at least one note.

But let's try to generalize the concept: if you want to know how many possible and different combinations of length "n" there are, with elements (including repeats) taken from a set of "m" possible choices, you simply need to *raise m to the power of n*.

Another example to illustrate this concept: take the letters of the English alphabet (m = 26). If you want to find out how many possible 3-character words can be formed with those letters (words here meant as random combinations of letters, thus including nonsensical ones), you need to calculate $26^3 = 17,576$ combinations, which, if we wanted to arrange them alphabetically, would range from "aaa" to "zzz."

If they were 5 letters, we would have had instead $26^5 = 11,881,376$ possible words.

And it is precisely for this reason that we are often asked to devise passwords for our web accounts that are composed of uppercase and lowercase characters, digits, and symbols, as well as being as long as possible: the fact that the password is made up of a wide range of elements increases m, while lengthening the string increases n. This way, the combinations needed for malicious software to decipher the password increase, making it impossible to uncover it even with thousands of years and computers with astronomical computing capabilities. Clearly, the vulnerability of a password also depends on concepts more complex than the simple "m and n," such as its amount of "entropy," meaning its internal disorder and complexity, but for now, let's set aside the details and stay within the realm of what is explorable through "pure" combination calculations.

Let's move on to another example: if you find yourself with a deck of playing cards and want to know how many different possible arrangements a shuffled deck can have, the problem is slightly different. Here, "n" (the length of the combination) is fixed and equal to "m" (the number of elements), and there are no possibilities for repetitions since, obviously, even if you shuffle them, the "cards are what they are." In this case, the operation to perform to find the number of permutations is different, and it is "n!." If you're not familiar with it, the exclamation mark means you have to multiply all positive integers from 1 up to n. But before performing this operation with the deck of cards, let's go through some simpler examples to better cement the concept.

If you have in your pocket two post-its, one with "a" written on it and one with "b," and you want to figure out how many different ways you can arrange them, first, you can note that n = 2. Then you calculate n!, which is 2! = 1 x 2 = 2. Obviously in fact, there are only two possible arrangements for two post-its: the arrangement "ab" and the arrangement "ba"! No other different arrangements are possible under these conditions.

If, instead, there were three pieces of paper with "a," "b," and "c" written on them, the arrangements would increase. For n = 3, you have n! = 3! = 1 x 2 x 3 = 6. In fact, you could arrange them as abc, acb, bca, bac, cab, cba.

But let's return to our deck of cards. In this case, you have n = 52, and thus, to find out all the possible arrangements, you will need to calculate 52! = 1 x 2 x 3 x 4 x 5 … x 50 x 51 x 52. This number is equal to

80.658.175.170.943.878.571.660.636.856.403.766.975.289.505.440.8
83.277.824.000.000.000.000

This means it is practically certain that no *sufficiently and truly shuffled* deck in the history of humanity has ever had the same order as another deck.

To be even clearer, imagine taking one hundred billion stars, each with a trillion planets, each planet with a trillion inhabitants, each inhabitant with a trillion decks of cards. If each card player could shuffle their deck a thousand times per second for 14 billion years (the supposed age of the Big Bang), only now would they begin to see the same permutations reappear, and it would take another 14 billion years of such an unrealistic scenario for them to repeat. This leads me to pose an "inevitable" provocation: if the deck of French cards in your drawer is likely unique in the entire history of the universe, *how unique are you truly as an individual?*

Save and feel better, with the 33% Rule

An extremely simple way to save on grocery costs might be the so-called "33% method." This method is based on the assumption that many people do not know (or tend to ignore out of laziness) that, on average, fresh products like fruits and vegetables cost much less per unit of weight compared to packaged products. A strategy to save on your weekly grocery shopping, therefore, could be to allocate 1/3 of your budget to fruit, 1/3 to vegetables, and the remaining 1/3 to everything else. So if, for example, your weekly grocery budget is 90 dollars, you could divide it as follows:

- 30 dollars for the vegetables
- 30 dollars for the fruit
- 30 dollars for everything else (basic products like pasta, rice, bread, dairy, meat, etc.)

Furthermore, as I always say, "every healthy eating plan starts with what you choose to have at home," this approach inherently brings the obvious yet often underestimated advantage of encouraging a much more balanced and nutritious diet. Not to mention the added side effect of getting us used to more frequently keeping track of our weekly expenses, helping us identify where we could save more and make more informed purchases. Of course, this is a simplified principle, and you can always refine it further depending on factors such as your budget or chosen diet.

Managing chaos

Many have surely heard of at least one of these: the butterfly effect or chaos theory. But for those who might be unfamiliar, or who desire a brief "refresher," let's start with some definitions: the term "butterfly effect" was coined by meteorologist Edward Lorenz, who observed that all weather models were extremely sensitive to initial conditions, to the point where weather forecasts could only be reliable over a very limited time period (as many of us know well). In this context, it seems that to illustrate just how sensitive such systems were, Lorenz himself suggested that the flap of a butterfly's wings in Brazil could potentially "unexpectedly" cause a tornado in Texas.

Well, the branch of mathematics that studies, through sophisticated equations, all systems that are extremely sensitive to initial conditions and, therefore, like weather systems, produce enormous differences as a result of very small changes, is called chaos theory.

In this book, we won't delve into anything particularly specific or detailed concerning this mathematical theory. For that, I recommend consulting far more authoritative manuals on the subject, such as *"Chaos: Making a New Science"* by James Gleick. However, I believe that by observing chaotic systems, even with the necessary simplifications, we can glean some interesting practical lessons that are extraordinarily useful for everyday life. For instance, if we were to approximate any real system over a sufficiently long period as if it were a chaotic system (meaning over time, "almost everything" inevitably tends to exhibit unpredictable perturbations compared to initial conditions), we might come to embrace what might at first seem intimidating but is, in my opinion, profoundly liberating – the principle of "letting go" of more things, more often. This doesn't mean neglecting our responsibilities or stopping our efforts to do our best, but rather embracing the awareness that we will never be able to predict or control how a "butterfly flap" which we might ignore today could, in five years, generate some unpredictable "tornado" (possibly a positive one; let's not turn this purely mathematical discourse into some "drama"). This could lead us to a powerful awakening, based on learning to live with a greater sense of curiosity and openness, and a lesser need to control every detail of our existence. Instead of overacting and overplanning in an attempt to craft a future that unfolds exactly as we wish, let's embrace a spirit of experimentation, continuous learning, and a conscious acceptance of the "chaos" that governs the mathematical variables of our universe!

And where, in my opinion, the greatest benefit can be drawn is in not confusing this acceptance with a completely nihilistic renunciation of the unknown. An ultimate lesson that we could indeed extract from this set of principles is to never underestimate the impact of a small action, a small bet, a small "yes" said to an opportunity we normally wouldn't have considered; even without the expectation of always knowing exactly where it will lead or insisting that each of them manifest something significant. The mindset of "trying small things anyway and letting them grow," I believe, can literally allow us to "leverage the chaotic systems" of the universe, opening us up to unprecedented opportunities for the revelation of extraordinary surprises.

- The Speed Math Bible -

Conclusions

Your journey into the world of mathematics, and rapid calculation comes to a close here, for now. I hope that despite the "tedium" that this subject instinctively communicates to many, I have managed the feat of making it at least a little more fascinating and enjoyable than you initially thought. But above all, I hope this book has provided you with the added value that every good book should offer its reader. In my opinion, this added value is found in being perceived not as an endpoint, but as the beginning of a journey toward intellectual goals. Hopefully, these goals are ever more ambitious and elevated.

I take this opportunity to thank you, the reader, for the affection you have shown for the "Kintsugi Project" through this purchase. And obviously… good luck with all your challenges!

Danilo Lapegna

But there's more...

Official English website, with our full catalog:

https://kintsugiproject.net/pages/welcome

Instagram:

@danilolapegna.kintsugi

Do you have any feedback for us? Suggestions? Advice? Write to us at info@kintsugiproject.net

Because the material we have to offer certainly doesn't end with this book. In fact, we sincerely hope that this volume is just the beginning of a wonderful journey together!

Danilo Lapegna

THE *CREATIVITY* CODE

SCIENCE-BASED METHODS FOR GENERATING BRILLIANT IDEAS.

The Kintsugi Project

- The Speed Math Bible -

Danilo Lapegna

NEURO HACKING

NATURAL METHODS TO BOOST ENERGY, SLEEP DEEPLY, REDUCE STRESS AND LEARN FASTER.

The Kintsugi Project

- The Speed Math Bible -

Danilo Lapegna

RESET!

HOW TO MASTER ART, SCIENCE AND PHILOSOPHY OF RADICAL, DURABLE CHANGE.

The Kintsugi Project

Danilo Lapegna

THE SCIENCE OF WINNING

HOW SCIENTIFIC METHOD CAN HELP US BECOME SMARTER, WEALTHIER, HAPPIER.

The Kintsugi Project

- *The Speed Math Bible -*

Danilo Lapegna

ULTRA HAPPINESS

A **COMPLETE, SCIENTIFIC GUIDE** FOR SPIRITUAL SURVIVAL IN THE MODERN WORLD.

The Kintsugi Project

- The Speed Math Bible -

The author

Danilo Lapegna, born in Italy in 1986, is the founder and CEO of the "Kintsugi Project." Based in Amsterdam, Netherlands, he is a tech project manager, a computer engineer, and an experienced writer with an insatiable passion for learning. From a young age, he showed an early fascination with the maximum potential of the human brain, devouring science-themed books and emerging as a television memory champion at the age of just six.

Thanks to his academic background in computer engineering, Danilo has successfully led international teams for years, working on high-impact software projects in the vibrant start-up scene of the United Kingdom. The complex management challenges within this highly competitive environment have fueled his growing passion and interest in a systemic and multidisciplinary approach to problems. This passion reaches its peak in his ability to generate value through rigorous data analysis and integration, driven by an unwavering desire to contribute to the well-being of others.

For over a decade, under the pseudonym "Yamada Takumi," he has leveraged his passions and skills to create a genuine "Scientific Quality Standard for Kintsugi," applied to books that have sold over 50,000 copies in Italy, climbing the sales charts on Amazon, helping thousands of people, and receiving enormous media attention for his success in the self-publishing industry.

And so, "The Kintsugi Project" represents the "ultimate" attempt by him and his staff to reinvent the approach to personal evolution, aiming to deconstruct all the "fluff" and the obsolete and dysfunctional paradigms of this sector. They then pivot their focus towards self-therapy systems, psychophysical well-being, "skill development," and "smart productivity" rooted in science, research, and above all, in a shared ecosystem that promotes individual and "personalized" growth, tailored to the values and needs of each person.

Bibliography and Further Reading

"Theory of Games and Economic Behavior" by John Von Neumann, Oskar Morgenstern (1944)

"Vedic Mathematics" by Bharati Krsna Tirthaji (1965)

"Game Theory: A Nontechnical Introduction" by Morton D. Davis (1970)

"A Course in Probability Theory" by Kai Lai Chung (1974)

"The Selfish Gene" by Richard Dawkins (1976)

"Short-Cut Math" by Gerard W. Kelly (1984)

"How Math Can Save Your Life" by James D. Stein (2013)

Appendix: Table of Prime Numbers from 2 to 5000

2 3 5 7 11 13 17 19 23
29 31 37 41 43 47 53 59 61 67
71 73 79 83 89 97 101 103 107 109
113 127 131 137 139 149 151 157 163 167
173 179 181 191 193 197 199 211 223 227
229 233 239 241 251 257 263 269 271 277
281 283 293 307 311 313 317 331 337 347
349 353 359 367 373 379 383 389 397 401
409 419 421 431 433 439 443 449 457 461
463 467 479 487 491 499 503 509 521 523
541 547 557 563 569 571 577 587 593 599
601 607 613 617 619 631 641 643 647 653
659 661 673 677 683 691 701 709 719 727
733 739 743 751 757 761 769 773 787 797
809 811 821 823 827 829 839 853 857 859
863 877 881 883 887 907 911 919 929 937
941 947 953 967 971 977 983 991 997 1009
1013 1019 1021 1031 1033 1039 1049 1051 1061 1063
1069 1087 1091 1093 1097 1103 1109 1117 1123 1129
1151 1153 1163 1171 1181 1187 1193 1201 1213 1217
1223 1229 1231 1237 1249 1259 1277 1279 1283 1289
1291 1297 1301 1303 1307 1319 1321 1327 1361 1367

1373 1381 1399 1409 1423 1427 1429 1433 1439 1447
1451 1453 1459 1471 1481 1483 1487 1489 1493 1499
1511 1523 1531 1543 1549 1553 1559 1567 1571 1579
1583 1597 1601 1607 1609 1613 1619 1621 1627 1637
1657 1663 1667 1669 1693 1697 1699 1709 1721 1723
1733 1741 1747 1753 1759 1777 1783 1787 1789 1801
1811 1823 1831 1847 1861 1867 1871 1873 1877 1879
1889 1901 1907 1913 1931 1933 1949 1951 1973 1979
1987 1993 1997 1999 2003 2011 2017 2027 2029 2039
2053 2063 2069 2081 2083 2087 2089 2099 2111 2113
2129 2131 2137 2141 2143 2153 2161 2179 2203 2207
2213 2221 2237 2239 2243 2251 2267 2269 2273 2281
2287 2293 2297 2309 2311 2333 2339 2341 2347 2351
2357 2371 2377 2381 2383 2389 2393 2399 2411 2417
2423 2437 2441 2447 2459 2467 2473 2477 2503 2521
2531 2539 2543 2549 2551 2557 2579 2591 2593 2609
2617 2621 2633 2647 2657 2659 2663 2671 2677 2683
2687 2689 2693 2699 2707 2711 2713 2719 2729 2731
2741 2749 2753 2767 2777 2789 2791 2797 2801 2803
2819 2833 2837 2843 2851 2857 2861 2879 2887 2897
2903 2909 2917 2927 2939 2953 2957 2963 2969 2971
2999 3001 3011 3019 3023 3037 3041 3049 3061 3067
3079 3083 3089 3109 3119 3121 3137 3163 3167 3169
3181 3187 3191 3203 3209 3217 3221 3229 3251 3253
3257 3259 3271 3299 3301 3307 3313 3319 3323 3329
3331 3343 3347 3359 3361 3371 3373 3389 3391 3407

3413 3433 3449 3457 3461 3463 3467 3469 3491 3499
3511 3517 3527 3529 3533 3539 3541 3547 3557 3559
3571 3581 3583 3593 3607 3613 3617 3623 3631 3637
3643 3659 3671 3673 3677 3691 3697 3701 3709 3719
3727 3733 3739 3761 3767 3769 3779 3793 3797 3803
3821 3823 3833 3847 3851 3853 3863 3877 3881 3889
3907 3911 3917 3919 3923 3929 3931 3943 3947 3967
3989 4001 4003 4007 4013 4019 4021 4027 4049 4051
4057 4073 4079 4091 4093 4099 4111 4127 4129 4133
4139 4153 4157 4159 4177 4201 4211 4217 4219 4229
4231 4241 4243 4253 4259 4261 4271 4273 4283 4289
4297 4327 4337 4339 4349 4357 4363 4373 4391 4397
4409 4421 4423 4441 4447 4451 4457 4463 4481 4483
4493 4507 4513 4517 4519 4523 4547 4549 4561 4567
4583 4591 4597 4603 4621 4637 4639 4643 4649 4651
4657 4663 4673 4679 4691 4703 4721 4723 4729 4733
4751 4759 4783 4787 4789 4793 4799 4801 4813 4817
4831 4861 4871 4877 4889 4903 4909 4919 4931 4933
4937 4943 4951 4957 4967 4969 4973 4987 4993 4999

- The Speed Math Bible -

Disclaimer

Any information, reference, or advice concerning psychological, psychotherapeutic, biological, or medical matters should in no way be considered a substitute for any type of practice with a qualified professional. The reader is fully responsible for what they do with the information contained in this book and is advised to consult professionals in the healthcare field before making any decisions that could potentially impact their health.

For any excerpt contained herein and cited from other works, the author appeals to the right of citation, the principles of fair use, and does not intend to economically harm any third-party authors or publishers in any way. Every content here belongs to its rightful owner, and the author encourages support, where possible, for the cited authors. For further information, please refer to the bibliography.

The scientific information included in this book is provided for educational and informational purposes. However, it should be noted that the authors assume limited responsibility regarding the accuracy, completeness, and timeliness of the information contained in the book. Science is an ever-evolving process, and some data may have already been replaced by new data or information. It is the reader's responsibility to verify the sources and information in the book and act accordingly.

In general, it is recommended that readers always use their own judgment and refer to the most up-to-date and reliable sources of information before making any decisions based on the information contained in any section of this book.

The authors hold all commercial and non-commercial rights to the visual and informational material included in this book, in all forms. Some images may have been generated using AI-based generative tools, but they have always been hand-finished through human, manual artistic work, which preserves all copyright rights for the "Kintsugi Project."

www.ingramcontent.com/pod-product-compliance
Lightning Source LLC
Chambersburg PA
CBHW071633220526
45469CB00002B/598